Schulenberg, Thomas
Starflinger, Jörg
(eds.)

High Performance Light Water Reactor

Design and Analyses

High Performance Light Water Reactor

Design and Analyses

Thomas Schulenberg
Jörg Starflinger
(eds.)

Impressum

Karlsruher Institut für Technologie (KIT)
KIT Scientific Publishing
Straße am Forum 2
D-76131 Karlsruhe
www.ksp.kit.edu

KIT – Universität des Landes Baden-Württemberg und nationales
Forschungszentrum in der Helmholtz-Gemeinschaft

KIT Scientific Publishing 2012
Print on Demand

ISBN 978-3-86644-817-9

Preface

There were not only us, the authors of this book, who wrote this book, produced results, tables and drawings, but a large team of scientists, nuclear engineers and students of nuclear energy who contributed with design, analyses and innovative ideas to the development of the High Performance Light Water Reactor. Three and a half years of joint development in the project "High Performance Light Water Reactor – Phase 2" in the 6th European Framework Program produced such a lot of results that the authors, who had a leading role in this project, felt an urgent need to summarize everything in a readable and consistent form such that later generations will have a chance to continue where we finished in 2010. In particular we like to mention the following partners of the team, in alphabetic order, who added quite significant contributions:

- Nusret Aksan, PSI, who advised us with safety system requirements,
- Michele Andreani, PSI, who managed the analyses of safety systems with different system codes,
- Henryk Anglart, KTH, who found methods to predict deteriorated heat transfer,
- Sylvie Aniel-Buchheit, CEA, who predicted the coolant temperature distribution in a fuel assembly with a sub-channel code,
- Olivier Antoni, CEA, who analyzed the performance of the safety system with CATHARE,
- Patrick Arnoux, CEA, who collected material data for structural analyses,
- Lars Behnke, KIT, who convinced us that dryout conditions at sub-critical pressure are not advisable, as part of his doctorate,
- Wolfgang Bernnat, IKE, University of Stuttgart, who predicted the change of the local power distribution in a fuel assembly cluster during burn-up,
- Dietmar Bittermann, AREVA, who integrated our various ideas to a reasonable power plant and thus managed the overall design concept,
- Martin Brandauer, KIT, who optimized the steam cycle and designed the preheaters as student of Karlsruhe University,
- Andreas Class, KIT, who found new methods to predict the flow and its stability limits in fuel assemblies,

- Andrea Conti, IKE, who performed Monte-Carlo-analyses of selected fuel assemblies as part of his doctorate at the University of Stuttgart,

- Antti Daavittila, VTT, who modeled a control rod ejection with TRAB-3D and SMABRE,

- Julien Drouin, KIT, who performed stress and deformation analyses of the backflow limiter as student of Karlsruhe University,

- Kai Fischer, EnBW, who designed the reactor pressure vessel as part of his doctorate at Karlsruhe University,

- Helena Foulon, KIT, who predicted flow and heat transfer of feedwater inside the reactor pressure vessel as student of Karlsruhe University,

- Eduard Guelton, who analyzed stresses of the reactor pressure vessel,

- Petr Hajek and Rudolph Vsolak, CVR, who designed and built a supercritical water loop for in-pile tests of cladding alloys,

- Liisa Heikinheimo, VTT, who managed the material test of different cladding alloys,

- Jochen Heinecke, AREVA, who challenged us to design the core with reasonable margins for hot channels and supported us with his experience in core design,

- Heiko Herbell, EnBW, who designed and analyzed the assembly box and later the reheater as part of his doctorate at Karlsruhe University,

- Steffen Himmel, KIT, who proposed the wire wrap as grid spacer and analyzed the flow inside assemblies with sub-channel analyses as part of his doctorate,

- Jan Hofmeister, RWE, who designed the fuel assembly by minimizing the structural material as part of his doctorate at University of Stuttgart,

- Attila Kiss, Budapest University, who predicted the detailed flow structure in a fuel assembly with CFD codes,

- Christina Köhly, KIT, who modeled the power plant and many of its components in the CAD-system CATIA,

- Claus Kunik, KIT, who predicted the flow structure in the gaps between assemblies as student of Karlsruhe University,

- Jona Kurki, VTT, who introduced a successful model for supercritical water in the system code APROS,

- Eckart Laurien, IKE, University of Stuttgart, who showed us how to use CFD codes properly for a flow of supercritical water,

- David Lemasson, who designed the feedwater tank during an internship at KIT,

- Matthias Löwenberg, RWE, who collected and prepared heat transfer data of supercritical water as part of his doctorate at University of Stuttgart,
- Jan-Aiso Lycklama à Nijeholt, NRG, who managed predictions of heat transfer in the fuel bundle,
- Marco Maisch and Steffen Siegel, KIT, who optimized and analyzed the concrete structure of the containment,
- Annalisa Manera, PSI, who analyzed the performance of the safety systems with RELAP5,
- Csaba Maráczy, KFKI, who predicted the core power distribution, together with Gy. Hegyi, G. Hordósy and E. Temesvári,
- Boudouin Tandeau de Marsac, who proposed the active and passive safety systems as student of Karlsruhe University
- Philippe Marsault, CEA, who managed design and analyses at CEA,
- Werner Meier, AREVA, who told us how to design a nuclear reactor core more realistically,
- Gaél Millet, who predicted stresses and deformations of the pressure vessel during an internship at KIT,
- Aurelien Miotto, who analyzed the flow structure of moderator water in the radial reflector as student of Karlsruhe University,
- Matthias Möbius, who analyzed the flow structure in the upper mixing chamber as student of Karlsruhe University,
- Lanfranco Monti, KIT, who coupled 4 different codes to predict the core power and density distribution pin-wise and locally as his doctorate,
- Radek Novotny, JRC, who performed cladding material tests in Petten,
- Tino Ortega Gomez, KIT, who predicted stability limits of coolant flow in the core, supported by R.T. Lahey,
- David Palko, KTH Stockholm, who was the first to predict the deteriorated heat transfer with a CFD code,
- Sami Penttilä and Aki Toivonen, who performed cladding material tests at VTT,
- Thomas Redon, who analyzed deformations and stresses of the steam plenum as student of Karlsruhe University,
- Tibor Reiss, who predicted stresses and deformations of pressure vessel and assembly boxes at KIT, as well as Xenon oscillations later at University of Budapest,

- Gerald Rimpault, CEA, who studied an alternative version of the core with a fast neutron spectrum, for comparison,

- Martin Rohde, University of Delft, who performed test of the stability limit with a surrogate fluid,

- Stephan Rothschmitt, AREVA, who designed the in-core instrumentation with C. Köhly,

- Mariane Ruzickova, CVR, who managed the in-pile corrosion and radiolysis tests of cladding alloys,

- Mala Seppälä, VTT, who predicted the performance of the safety system with SMABRE,

- Marc Schlagenhaufer, KIT, who designed and analyzed the control of the power plant and its safety system with APROS, as his doctorate,

- Tobias Schlageter, KIT, who analyzed stresses and deformations of a tube sheet of a preheater as a student of Karlsruhe University,

- Tobias Schneider, who designed the foot piece of the fuel assembly cluster as a student of Karlsruhe University,

- Medhat Sharabi, PSI, who made analyses with RELAP investigating the effect of heat transfer correlations during steady-state and transient operation,

- Xavier Tiret, who modeled flow reversal phenomena in the first superheater as student of Karlsruhe University,

- Timo Vanttola, who managed system analyses at VTT,

- Rémi Velluet, who designed the separators during an internship at KIT,

- Dirk Visser and Laltu Chandra, NRG, who predicted the local flow phenomena and heat transfer near wire wraps of fuel rods,

- Bastian Vogt, EnBW, who proposed an evolutionary approach from PWR to HPLWR as his doctorate at KIT,

- Peter Volkholz, AREVA, who supported us designing the safety system,

- Christine Waata, KIT, who performed the first coupled analysis of flow and power distribution in a HPLWR fuel assembly as her doctorate,

- Alexander Wank, KIT, who optimized coolant mixing in the upper and lower mixing chambers as his doctorate,

- Michael Wechsung, Siemens, who designed the steam turbine concept for the HPLWR,

- Maxime Werner, KIT, who optimized valves and orifices of the safety system as student of Karlsruhe University,

- Yu Zhu, IKE, who analyzed the flow in fuel assemblies as his doctorate at University of Stuttgart.

The authors like to thank the European Commission for their financial support of the HPLWR Phase 2 Project under the contract FI6O-036230, which enabled this successful European collaboration, and to thank all other sponsors of the project, in particular EnBW Kernkraft GmbH, RWE Power AG and AREVA NP GmbH, who contributed additional grants for students and doctorates. We also like to thank the members of our advisory board

- Yoshiaki Oka, Tokyo University, who contributed with guidance through the international SCWR developments

- Oddbjörn Sandervåg, MME, who provided us the regulators point of view of our very innovative system

- Liisa Heikinheimo, TVO, who later represented a utility in our advisory group,

and

- Georges von Goethem, EU, who informed us about other system and development projects also co-sponsored by the European Commission currently being investigated within Generation IV International Forum.

Last not least, we like to thank all partners in Japan, Canada and South Korea, who helped us through information exchange and discussion in the Generation IV International Forum to work out this innovative concept.

About the Authors

Prof. Dr.-Ing. Thomas Schulenberg is head of the Institute for Nuclear and Energy Technologies at Karlsruhe Institute of Technology, Germany. He was managing the working package "Core Design" in the European project HPLWR-Phase 2 and he is representing Euratom member states in the Steering Committee for Supercritical Water Cooled Reactors of the Generation IV International Forum.

Prof. Dr.-Ing Jörg Starflinger is head of the Institute of Nuclear Technology and Energy Systems at the University of Stuttgart, Germany. He has been coordinator of the European project HPLWR-Phase 2 and he is member of the Steering Committee for Supercritical Water Cooled Reactors of the Generation IV International Forum.

Contents

1   Introduction 1
1.1   Past experience 2
1.2   New recent approaches 9
1.2.1   The first HPLWR project 9
1.2.2   Optimization of fuel assembly design 13
1.2.3   Core design concepts 20
1.2.4   Phase 2 of the HPLWR project 27
1.3   References 28
2   Core design 33
2.1   Design target 33
2.2   General design strategy 34
2.3   Mechanical design 37
2.4   Analysis of the core power distribution 43
2.5   Local peaking factors 56
2.6   Coolant mixing 60
2.7   Moderator flow 66
2.8   Stability issues 68
2.8.1   Density wave oscillations 68
2.8.2   Flow reversal 71
2.8.3   Xenon oscillations 74
2.9   Stresses and deformations of the assembly box 76
2.10   Sealing against by-pass flows 80
2.11   Uncertainties of the analyses 81
2.12   Hot channel assessment 84
2.13   Discussion 86
2.14   References 87
3   Primary System 93
3.1   General design strategy 93
3.2   Design of the reactor pressure vessel and its internals 95
3.3   Analysis of the feedwater flowpath 99
3.3.1   Backflow limiter 99
3.3.2   Moderator flowpath 102
3.3.3   Downcomer flow 105
3.4   Stress analyses 107
3.4.1   Reactor pressure vessel 107
3.4.2   Reactor pressure vessel internals 116
3.5   In-core instrumentation 118
3.6   Fluence analysis 123
3.7   Discussion 125
3.8   References 126
4   Containment and Safety Systems 129
4.1   General strategy of the safety system 132
4.2   Design of the containment and its safety systems 140
4.3   System response to postulated accidents 145
4.3.1   Simulation of the main safety LPCI system 150

4.3.2 Transients addressing heat storage capacity 153
4.3.3 Transients addressing core cooling in case of loss of flow 155
4.3.4 Core coolability in case of loss of coolant 160
4.3.5 Consequences of a control rod ejection 168
4.3.6 Other events 173
4.3.7 Consequences for safety system design 174
4.4 Mechanical analyses of the containment 177
4.5 Discussion 181
4.6 References 183
5 Steam Cycle and Layout of the Power Plant 185
5.1 Design strategy 186
5.2 Thermodynamic options 188
5.3 Major components of the steam cycle 191
5.3.1 Steam turbines 191
5.3.2 Reheater 194
5.3.3 Feedwater pumps 197
5.3.4 Feedwater tank 198
5.3.5 Feedwater preheaters 199
5.4 Control of the steam cycle 203
5.4.1 Control in the load range 203
5.4.2 Start-up and shut-down procedure 205
5.5 Layout of the power plant 214
5.6 Cost estimate and discussion 219
5.7 References 226
A Annex 1 229
HPLWR Design Data 229
Overall Performance 230
Balance of Plant 230
Nuclear island (inside containment) 233
Reactor Pressure Vessel and Internals 236
Core and Fuel Assemblies 239

1   Introduction

Looking at the trend of coal fired power plants in the last 40 years, Fig. 1.1, we observe a remarkable increase of net efficiency from around 37% in the 1970ies to more than 46% today. The last 20 years since 1990, in particular, were characterized by an increase of live steam temperature beyond 550°C, when boiler steels became available which allowed to exceed the former material limits. Along with the temperature increase, the live steam pressure went up to maximize the turbine power, finally exceeding the critical pressure of water. The next generation of coal fired power plant will even reach a net efficiency of ~50%, when live steam temperatures of 700°C or more can be realized. In comparison with such development, the net efficiency of latest pressurized water reactors (PWR) of around 36% is still close to the efficiency of ~34% of the first generation of light water reactors.

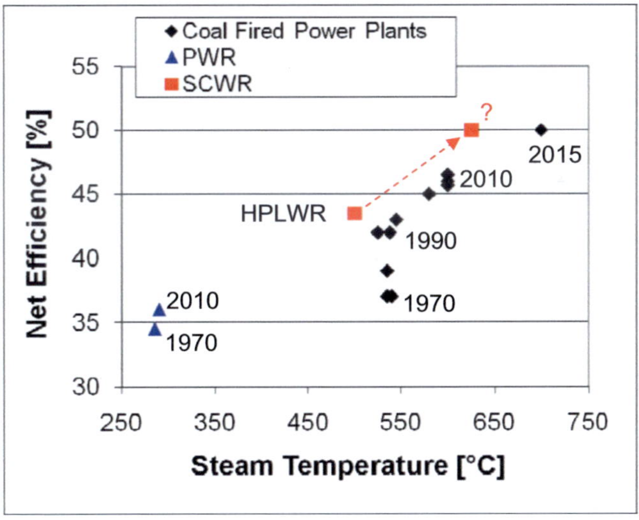

Fig. 1.1 Increase of the net efficiency with live steam temperature of different thermal power plants

This comparison had motivated the Generation IV International Forum to look for similar options for future light water reactors with superheated steam at supercritical pressure, called Supercritical Water Cooled Reactors (SCWR). The higher steam enthalpy could enable a direct, once through steam cycle such that neither steam generators nor steam separators and dryers would be required, and even primary coolant pumps could be omitted. Moreover, steam turbines and reheaters could be significantly smaller than today. The power plant, therefore, should have less plant erection costs, while the thermal efficiency would even be higher – a clear economic advantage. As supercritical water does not boil with a phase change, a boiling crisis would physically be excluded, and superheated steam could be produced without excessive hot spots in the core. Moreover, as fossil fired power plants with supercritical steam conditions have been operated since 20 years now, this nuclear plant concept could benefit from proven design of turbines, feedwater pumps and most other components of the steam cycle, except the reheater. On the other hand, the containment design can basically be derived from latest boiling water reactors (BWR), so that the research and development program would need to concentrate mainly on the reactor itself.

1.1 Past experience

Supercritical water cooled reactors were studied already in the 1950ies and 1960ies as summarized by Oka [1]. In particular, we like to mention the following early studies:

- A light water moderated, supercritical steam cooled reactor was designed by Westinghouse in 1957, in which 7 fuel rods each in cylindrical, double walled cans formed the fuel assemblies to insulated the superheated steam from the liquid moderator water at 260°C. An indirect steam cycle was favored for this concept to avoid activity in the turbines.

- A heavy water moderated reactor, cooled with light water, was designed by General Electric in 1959 for a thermal power of 300 MW with a once through steam cycle. The coolant was passing the core four times, reaching an outlet temperature of 621°C.

- A graphite moderated and light water cooled pressure tube reactor was designed by Westinghouse in 1962, called the Supercritical Once Through Tube Reactor (SCOTT-R) for an electric power of 1000 MW with a thermal efficiency of 43.5%. The low pressure tank containing the graphite moderator was cooled with Helium.

- A pressurized water reactor with a closed loop primary system at supercritical pressure had already been proposed in 1966.

A supercritical water cooled reactor, however, has never been built in the past. Instead, a boiling water reactor with a nuclear superheater was built in Grosswelzheim, Germany, which could be considered as an early, evolutionary step from boiling water reactors towards an SCWR. The HDR (Heissdampfreaktor) by AEG was intended to reach 500°C core outlet temperature in its final stage, and the prototype built from 1965 to 1969 with 100 MW thermal power was designed for a reduced temperature of 457°C of superheated steam at 9

MPa reactor inlet pressure as an introductory step. Its characteristic data are listed in Tab. 1.1, taken from Dörfler [2] and from Traube and Seyfferth [3].

The reactor core consisted of 52 fuel assemblies with 24 fuel rods of annular cross section in a square assembly box with 178 mm outer width as sketched in Fig. 1.2. The outer claddings had an outer diameter of 26.5 mm at a pitch of 33.75 mm and provided 75 MW thermal power to evaporate the coolant between the fuel rods like in a conventional BWR. The inner cladding with an inner diameter of 12.5 mm served as the superheater which provided 25 MW thermal power to superheat the steam. Six of these fuel rods each, indicated with a dotted line in Fig. 1.2, were combined to a superheater loop through which the steam was driven four times through the core without intermediate mixing. Saturated steam was guided in a common steam tube per assembly from the upper plenum of the reactor, split into 4 tubes in a manifold at the top of the assembly box and led to the inside of fuel rods 1 and 3 of each six-pack where it was driven downwards. After a U-bend with a cross over underneath the core, as shown in Fig. 1.4 (right), the steam moved upwards again through the inside of fuel rods 2 and 4, respectively. Both steam lines were combined then to a single line flowing downwards through fuel rod 5 and upwards again through fuel rod 6. All steam lines of the 4 six-packs of each assembly were combined then in a header to a common superheater tube to run upwards to a steam plenum. The inner rod of the 5x5 rod assembly was filled with a Gadolinium rod inside a Zircalloy cladding as a burnable poison to compensate the excess reactivity.

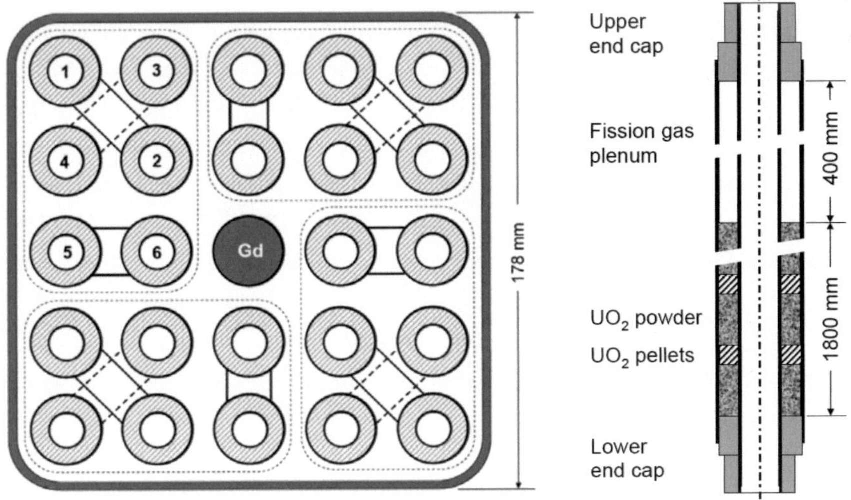

Fig. 1.2: Cross section of an HDR fuel assembly and sketch of the fuel rod design [4]

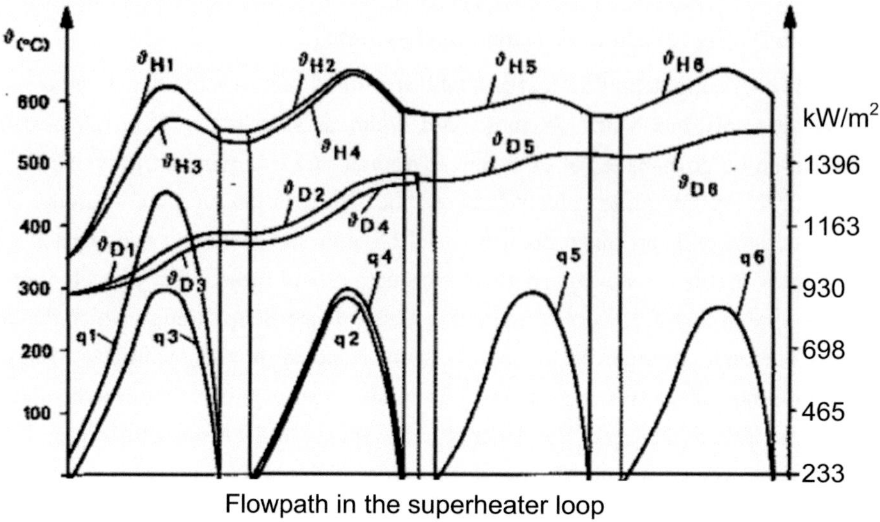

Fig. 1.3: Coolant temperatures ϑ_D, inner cladding temperatures ϑ_H and surface heat flux q of the hottest superheater loop in the core. Indices refer to fuel rod numbering in Fig. 1.2 [3]

The superheater loops caused a significant pressure drop such that the reactor outlet pressure (or live steam pressure) was only 7.33 MPa. Fig. 1.3 shows that the heat flux q of the inner superheater claddings 1 - 6 had a peak of up to 1300 kW/m² in the hottest fuel assembly, producing up to 550°C peak coolant temperature ϑ_D. The cladding temperatures ϑ_H of these fuel rods were reaching a peak temperature of up to 650°C. Therefore, the inner claddings had to be made from Inconel 625, whereas the outer claddings could be made from stainless steel 1.4981 S (X 8 CrNiMoNb 16 16) or 1.4550 (X 6 CrNiNb 18 10, equiv. to SS 347) as back-up material. The assembly box, on the other hand, was exposed only to the saturation temperature of around 300°C and could be made from Zircalloy 4 to minimize neutron absorption. The fuel was enriched UO_2 with a U-235 enrichment of 3.16% in all fuel rods except the corner rods of each assembly with 2.6% enrichment only. It was made from compacted UO_2 powder with sintered UO_2 pellets in between to align the inner cladding, as sketched in Fig. 1.2 (right). Cruciform neutron absorber rods were inserted from the core bottom and were running between the assembly boxes like in a conventional BWR, as indicated in Fig. 1.5.

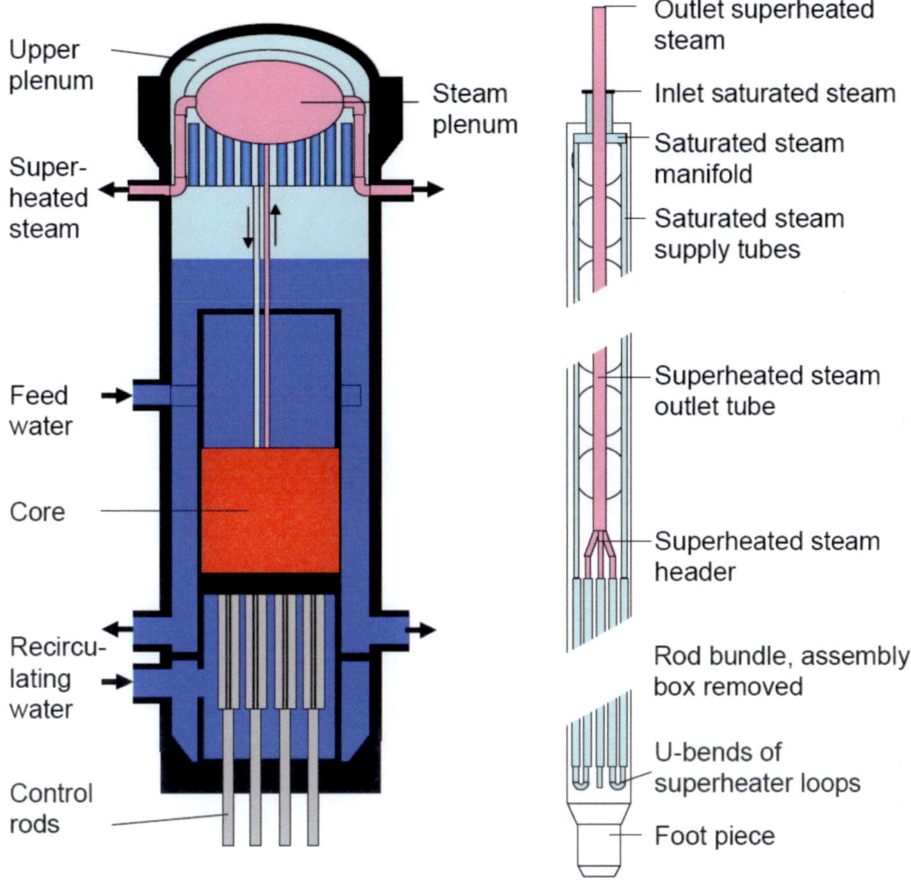

Fig. 1.4: Sketch of the HDR reactor design [3] and of a fuel assembly [4]. Liquid water shown in dark blue, saturated steam in light blue, and superheated steam in pink

Fig. 1.4 (left) shows a simplified sketch of the reactor. The core with its active fuel height of 1.8 m plus a fission gas plenum of 400 mm can be found in the lower half of the reactor. Two outside recirculation pumps provided a coolant mass flow of 644 kg/s through the core. Steam separators in the upper part of the reactor released saturated steam to the upper plenum, from where tubes guided the saturated steam through the upper assembly box to the 4 superheater loops in the core, as described above. The superheated steam was collected above the core and led through superheater tubes (one per assembly) to the steam plenum inside the upper plenum. Fig. 1.4 (right) shows a cut away view of a fuel assembly with a total length of 6835 mm.

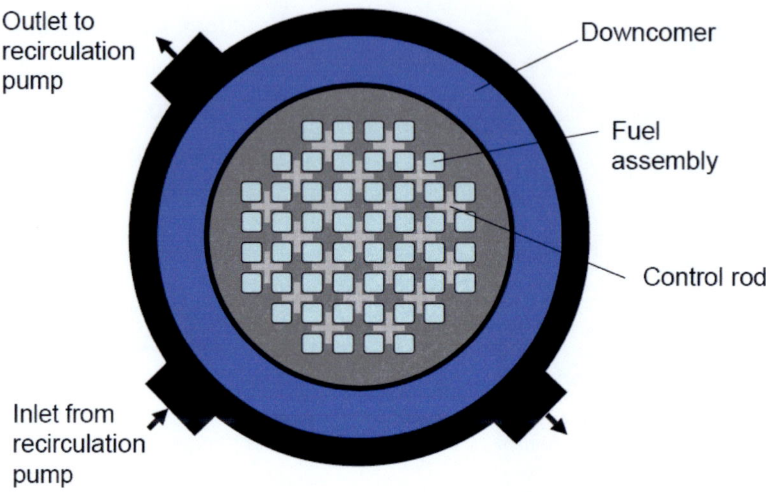

Fig. 1.5: Cross section of the reactor core with 52 fuel assemblies and 21 absorber elements as control rods [3]

The reactor produced 43.6 kg/s of superheated steam in total. As a suitable turbine for a direct steam cycle had not been available, a secondary steam cycle was built with an evaporator and superheater, using an older steam turbine from Dettingen power plant producing 25 MW electric power, as shown in Fig. 1.6. The secondary steam cycle was designed with a feedwater temperature of 110°C, a steam pressure of 3.5 MPa and a live steam temperature of 430°C at the superheater outlet. This steam had to be cooled down by feedwater injection to 390°C and expanded to 1.6 MPa to meet the turbine requirements. The gross thermal efficiency of 25% is thus not representative for this power plant but rather a pragmatic solution to produce electric power during the test phase.

The fuel assembly was tested out of pile and later in-pile in the research reactor Kahl as summarized by Höchel and Fricke [4]. In-pile test runs with thermal-hydraulic monitoring were performed up to 546 days with a maximum burn-up of 11.9 GWd/t. The maximum steam temperature of 520°C and the maximum cladding temperature of 650°C on the inner, superheater side were close to the expected reactor conditions. Damage of the inner cladding was found only occasionally in case of the hottest material temperatures of around 650°C. One of the test fuel assemblies was damaged because of a temporary reduction of the steam mass flow rate. Additional mechanical, thermal-hydraulic and emergency cooling tests of the fuel assembly demonstrate that the prototype design was well prepared.

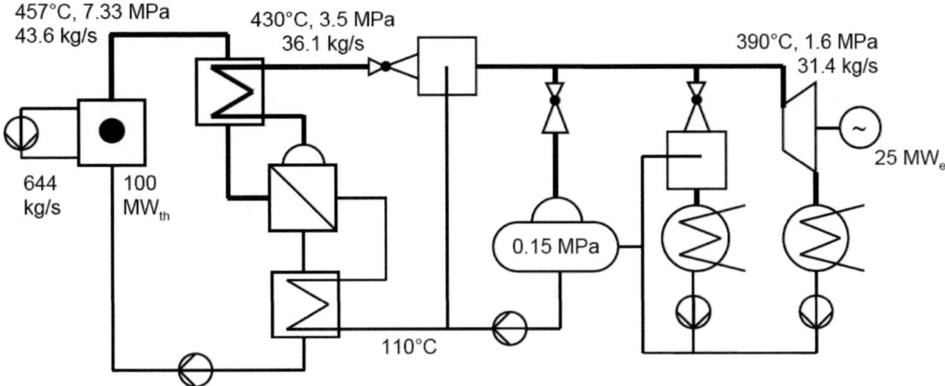

Fig. 1.6: Steam cycle of the HDR prototype power plant in Grosswelzheim [2]

The construction of the power plant started in Jan. 1965. First criticality was reached on Oct. 14, 1969, and the HDR power plant was connected with the electricity grid on the same day. Commercial operation started on Aug. 2, 1970, reaching up to 23 MW_e, but the core was damaged soon. We were told that the tubes of the superheated steam collapsed, but details have not been published. The reactor tests were finished and the reactor was shut down already on April 20, 1971, only 18 months later, having produced 6200 MWh electric power in total which is equivalent to about 10 full power days or to an average burn up of about 0.15 GWd/t.

The history of this power plant must also be regarded in connection with the development of AEG, who merged their nuclear division with the one of Siemens to KWU (Kraftwerk Union) in 1969. AEG had been producing BWRs based on General Electric technologies and Siemens had been building PWRs based on Westinghouse technologies. The conventional BWR of AEG built in Würgassen had been more reliable at that time and KWU decided to concentrate their further joint development on the conventional BWR, which became known as the BWR Type 69 and which was running successfully in 4 nuclear power plants in Germany until 2011. The HDR, on the other hand, was given up to reduce the product portfolio.

The HDR facility was used later for severe accident tests in the 1980ies and was finally pulled down in the 1990ies. Since 1998, there is nothing left anymore but a green field where the reactor once had been. Klaus Traube, one of the inventors of the HDR and co-author of [3] was leading afterwards the development of the sodium cooled fast breeder reactor SNR-300 at Interatom, Germany, until 1976, but changed to the anti-nuclear movement after he was falsely suspected of passing on secret information to people sympathizing with terrorism.

7

In total, therefore, not a story of success but still a milestone in the development of light water reactors with increased temperatures.

Primary system		
Total thermal power	100	MW
Thermal power of the evaporator	75	MW
Thermal power of the superheater	25	MW
Pressure at reactor inlet	9	MPa
Superheated steam temperature	457	°C
Superheated steam pressure at reactor outlet	7.33	MPa
Mass flow of superheated steam	43.6	kg/s
Evaporator mass flow rate	644	kg/s
Number of fuel assemblies	52	
Number of fuel rods per assembly	24	
Number of Gd rods per assembly	1 (central)	
Active core height	1800	mm
Fission gas plenum	400	mm
Outer diameter of the outer fuel cladding	26.5	mm
Wall thickness of outer cladding	0.6	mm
Material of outer cladding	1.4981 S	
Inner diameter of the inner cladding	12.5	mm
Wall thickness of inner cladding	0.5	mm
Material of inner cladding	Inconel 625	
Pitch of fuel rods	33.75	mm
UO_2 enrichment except corner rods	3.16	%
UO_2 enrichment of corner rods	2.6	%
Design peak temperature of inner cladding	650	°C
Outer width of assembly box	178	mm
Material of assembly box	Zircalloy 4	
Number of absorber elements	21	
Secondary system		
Turbine power	25	MW
Steam pressure	3.5	MPa
Steam temperature	430	°C
Feedwater temperature	110	°C
Steam mass flow rate	36.1	kg/s

Tab. 1.1: Characteristic data of the HDR prototype in Grosswelzheim, Germany, with superheated steam [2], [3]

1.2   New recent approaches

A new approach was taken in the 1990ies in Japan, where Y. Oka started to work again on SCWR concepts with a group of scientists and students at the University of Tokyo, first concentrating on a core with a thermal neutron spectrum and later also on an alternative concept with a fast neutron spectrum. One of their first concepts to mention is a thermal reactor design published by Dobashi et al. [5] with hexagonal fuel assemblies with round water rods, which were thermally insulated against the superheated steam by stagnant water layers. The concept was improved later by Yamaji et al. [6] using square fuel assemblies with 300 fuel rods and with 36 square water rods through which feedwater was running downwards. The coolant was running downwards as well and was thus preheated in the outer, peripheral fuel assemblies of the core. It was mixed then with the moderator water underneath the core and was finally heated up to 500°C with an upward flow through the central fuel assemblies of the core. The fast reactor design concept of Cao et al. [7] was using a similar flow path but hexagonal fuel assemblies with a tighter lattice instead. Water rods for moderation were omitted and a negative void coefficient was reached after some core design optimization by mixing seed assemblies with blanket assemblies with solid ZrH moderator layers. All these design concepts and their analyses have been documented well in a recent book of Oka et al. [8], so that we can concentrate rather on the development in Europe instead.

1.2.1   The first HPLWR project

It was in 2000 that a consortium of European research institutes and industrial partners got interested in the Japanese design studies so that they agreed to take a closer look at the potential merits and plant characteristics of a SCWR and to indentify the major issues developing it. Co-sponsored by the European Commission in the 5th Framework Program and coordinated by D. Squarer, they launched a 2 years project called "High Performance Light Water Reactor" (HPLWR), avoiding the term "supercritical" which could be misunderstood when talking about nuclear reactors (it is still a critical reactor neutronically). A summary of the project is given by Squarer et al. [9]. The team took advice from Y. Oka who had already 10 years of experience with SCWR by that time. A preliminary selection was made for the HPLWR scale, for boundary conditions, for core and fuel assembly design, as well as for reactor pressure vessel, containment, turbine and balance of plant design. Potential materials were reviewed and selected, and codes for safety analyses were started to be upgraded for supercritical water conditions. The following sections are summarizing the most important results of this project.

The potential plant characteristics or, better to say, reasonable design targets worked out in the HPLWR project, are listed in Tab. 1.2. The electric power of 1000 MW was targeted on base load power producers without using yet the economic advantages of a larger scale. The system pressure of 25 MPa and the core inlet and outlet temperatures of 280°C and 500°C, resp., were following the proposal of Dobashi et al. [5], whereas the active core height of 4.2 m and the outer fuel pin diameter of 8 mm were typical target data of advanced pressurized water reactors. Target data for burn-up, design life, costs or safety features were driven by latest light water reactor development which needed to be exceeded by the present study to remain to be competitive.

Net electric power	1000	MW
Feedwater temperature	280	°C
Steam temperature	500	°C
System pressure	25	MPa
Envisaged advantage of plant erection costs	20 to 25	%
Maximum burn-up target	60	GWd/t
Design life	60	Years
Core damage frequency target	$< 10^{-5}$	per year
Severe accident release target	$< 10^{-6}$	per year

Tab. 1.2: General design target of the High Performance Light Water Reactor [10].

The assembly design concepts which were studied were initially based on proposals of Dobashi et al. [5], but alternative concepts were published also by Bittermann et al. in [10]: In order to improve moderation in spite of the low coolant density of superheated steam, they proposed to increase the number of water rods in the assembly as indicated in Fig. 1.7 or even to include solid ZrH rods as solid moderator. The first reactor pressure vessel design, Fig. 1.8, was solving already the problem of thermal stresses and larger deformations to be expected if the thick walled structure were exposed to feedwater and to superheated steam: contact of steam with the pressure vessel was excluded by a coaxial supply of feedwater around the hot steam tube, shown of the right hand side. The safety system design concept was basically taken from the boiling water reactor design SWR 1000, as sketched in Fig. 1.9, with a pressure suppression pool, low pressure coolant injection, core flooding pools, containment condensers and emergency condensers. The steam cycle, sketched in the same figure, was merging the conventional BWR concept with a start-up system taken from supercritical fossil fired power plants with sliding pressure operation.

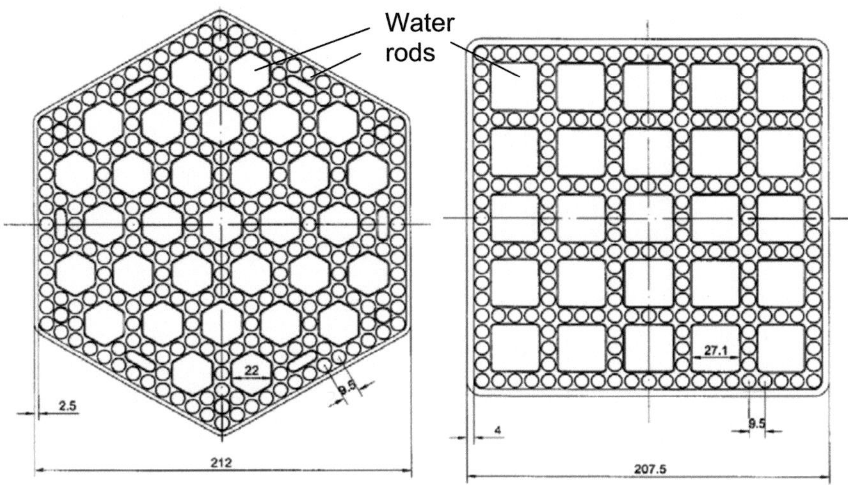

Fig. 1.7: Early assembly design concepts with increased moderation ratio proposed in the HPLWR project [10]; left: hexagonal arrangement, right: square arrangement (type sq1.5)

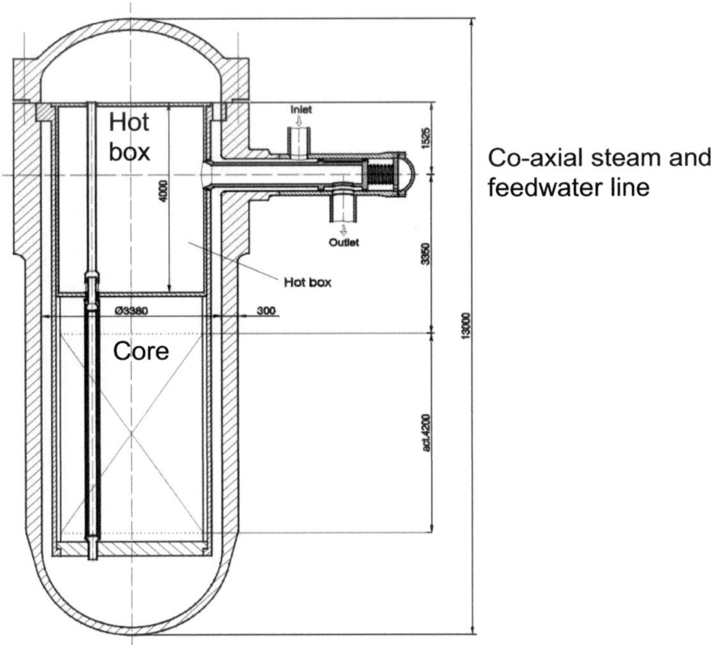

Fig. 1.8: One of the first design sketches of the HPLWR pressure vessel [10]

11

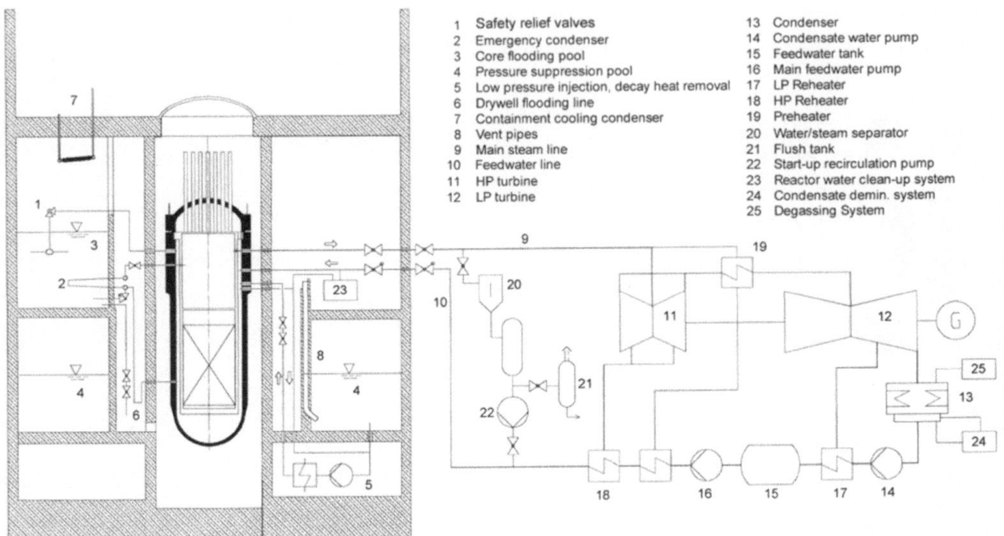

Fig. 1.9. Sketch of the HPLWR safety systems and steam cycle as proposed in [10]

Potential cladding materials for fuel rods of the HPLWR were discussed by Ehrlich et al. [11]. They identified early that stainless steels could be a reasonable compromise between the creep and corrosion requirements at temperatures beyond 600°C, on one hand, and neutronic requirements for minimum neutron absorption raised by Rimpault et al. [12] on the other hand.

A first impression on safety system behavior in case of loss of coolant accidents could be obtained with transient system analyses of Antoni and Dumaz [13], who extended successfully the core CATHARE to supercritical water conditions to model the consequences of a break of the feedwater line and of the steam line. While a break of the steam line would rather overcool the core temporarily because of the increased coolant mass flow during depressurization, they noticed a hot peak of the cladding temperature in case of a break of the feedwater line. This was a first indication that the system performance will be different from a boiling water reactor and the concept of the safety system cannot just be copied from there.

Even though none of the reactor or system components had been designed in detail yet, the HPLWR project was forming a solid basis for future studies and most of the boundary conditions and design targets defined there have been kept within the following 10 years.

1.2.2   Optimization of fuel assembly design

Work on the detailed design started first with an optimization of the fuel assembly. As a first step, Cheng et al. [14] took the design proposals shown in Fig. 1.7 and optimized the pitch to diameter ratio of the fuel rods for hexagonal and square arrangements. A sub-channel code STAFAS was written including heat transfer correlations for supercritical water at the fuel claddings as well as heat transfer through the assembly or water boxes to the moderator water. With an outer diameter of 8 mm of the fuel claddings and a duct wall clearance of 1 mm, they predicted the lowest cladding temperatures with a pitch to diameter ratio of 1.15 in case of square assemblies and of 1.3 in case of hexagonal assemblies. A uniform power production was assumed for all fuel rods in this purely thermal-hydraulic study, except for the corner rods which were assumed to be dummy rods.

As a second step, Hofmeister et al. [[15], [16]], optimized the size of the fuel assemblies and thus the number of fuel rods per assembly with simple, algebraic analyses to get a uniform neutron moderation and uniform coolant heat up, a high power density and a minimum of structural material in the core, which turned out to be the key problems for core design in the study of Rimpault et al. [12]. They estimated the required wall thickness s of an assembly box with side length l under a pressure load Δp as

$$s = \sqrt[3]{\frac{\Delta p \; l^4}{32 \; E \; f}}$$

where E is the Young's modulus and f is a given deflection of the box wall which may not be exceeded to avoid flow blockage inside. This estimation allowed predicting the ratio of structural material to fuel, Fig. 1.12, with the result that assembly boxes with larger width, like in Fig. 1.7, need a higher ratio than smaller ones, hexagonal assemblies, like in Fig. 1.11, have a smaller ratio than square ones, like in Fig. 1.10 and, of course, less water boxes per fuel rod decrease this ratio. As neutronic analyses of Rimpault et al. [12] indicated that a fuel rod should always have a box wall and thus moderator water in its neighborhood, they allowed only two rows of fuel rods between two parallel box walls at maximum, like sq2.1 and hex2.1 in Fig. 1.10 and 1.11 (right).

Looking at the ratio of moderator plus coolant to fuel at different axial height in the core, Fig. 1.13, the square assemblies were superior to the hexagonal ones, and a small square assembly with 40 rods and a single water box in its center, Fig. 1.10 (right) came close to the moderator to fuel ratio as a pressurized water reactor (PWR).

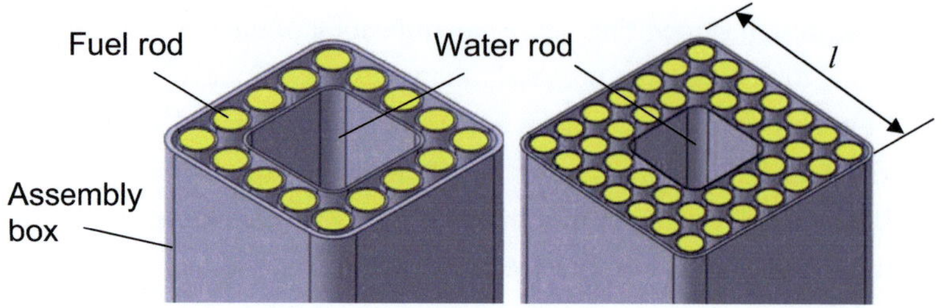

Fig. 1.10: Small size square fuel assembly concepts sq1.1 and sq2.1 [19]

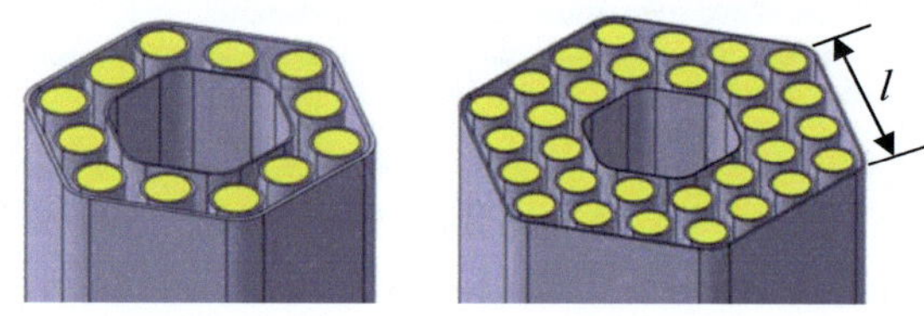

Fig. 1.11: Small size hexagonal fuel assemblies hex1.1 and hex 2.1 [19]

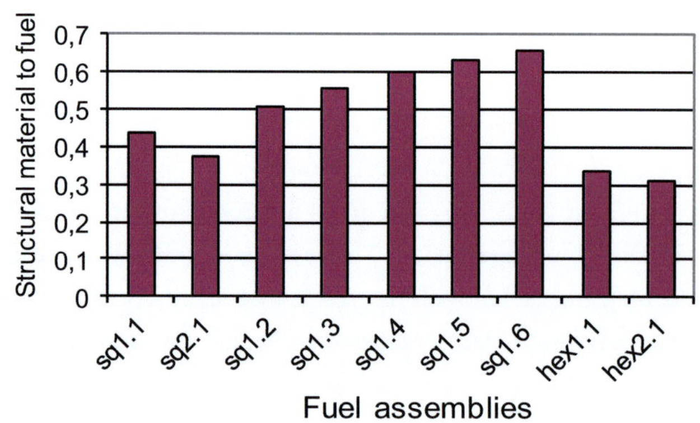

Fig. 1.12 Ratio of structural material to fuel mass of different assembly concepts [15]

A simplified analyses was performed also to estimate the non-uniformity of coolant heat up, neglecting any coolant mixing between sub-channels, which showed again some advantage of the square assembly, so that finally the assembly sq2.1 shown in Fig. 1.10 (right) was selected as best.

Waata et al. [[17], [18]], coupled the sub-channel code STAFAS [14] with the Monte-Carlo neutron physics code MCNP to predict more precisely the power, temperature and coolant density distribution in this assembly. Moderator water was assumed to flow downwards in gaps between assembly boxes and in the central water rod. It was heated up by the hotter box walls, was assumed to mix with more feedwater underneath the fuel assembly and finally rising upwards as coolant between the fuel rods. Reflecting boundary conditions for all neutrons were assumed in the MCNP analysis, simulating 1/8 of a single assembly as a representative geometry. An enrichment of 5% was assumed for all fuel rods except the corner rod which had 4% only [18]. The moderator water was assumed to enter with 280°C at the top and the coolant with 300°C at the bottom of the assembly. Results for the coolant temperature of this single assembly analysis are shown exemplarily in Fig. 1.14. Due to heat losses through the inner and outer box walls, the central sub-channels 3, 4 and 7 became hotter than those ones next to a box wall, which indicated the need for a thermal insulation of the box walls.

A summary of this joint optimization study has been published by Hofmeister et al. [19].

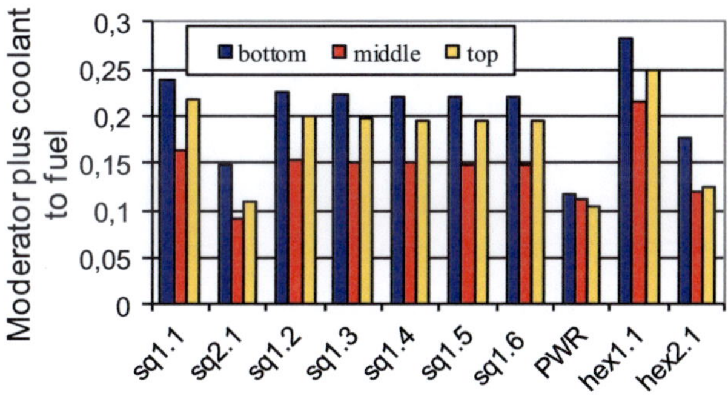

Fig. 1.13: Ratio of moderator plus coolant to fuel mass of different assembly concepts [15]

An approach to enhance heat transfer and mixing in this fuel assembly has been taken by Bastron et al. [20] and Meyer et al. [21]. Heat transfer can be enhanced at least by a factor of 2 with an artificial surface roughness increasing turbulence. For a fuel rod of 8 mm diameter, they recommend circumferential ribs of about 0.12 mm height. They biggest effect with an enhancement factor of 2.5 is expected at temperatures around the pseudo-critical point at 384°C, i.e. the range below around 2 m axial height in Fig. 1.14. However, the friction losses due to these ribs will increase by a factor of 8 compared with a smooth rod. Therefore, at higher coolant temperatures and thus lower coolant density, the ribs should rather be avoided as their effect on heat transfer is small then, but the friction losses are highest there.

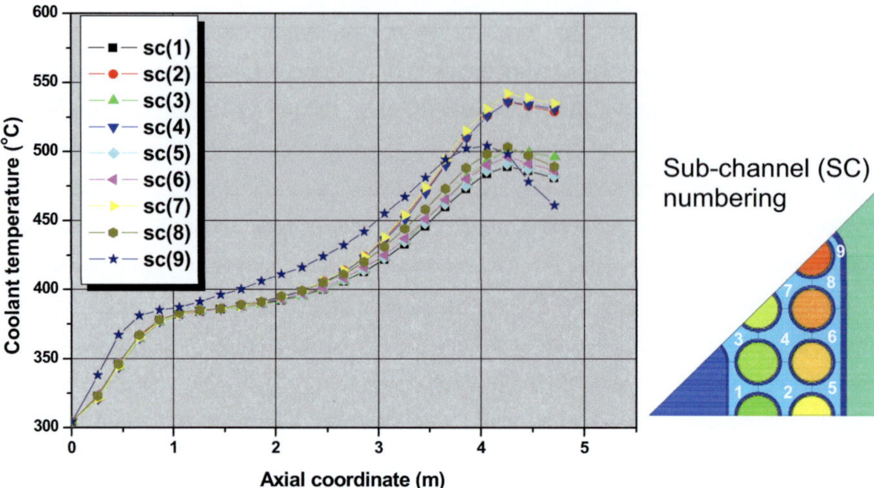

Fig. 1.14: Coolant temperature distribution of a single fuel assembly with an average coolant outlet temperature of 508°C; coupled MCNP-STAFAS analysis [18]

A similar effect on heat transfer and simultaneously an improved mixing between sub-channels was predicted by Bastron et al. [[20], [21]], for staircase grid spacers as shown in Fig. 1.15. By partially blocking the coolant flow path, this innovative spacer design will cause a swirl flow inside the assembly with an enhancement of heat transfer of a factor of around 2, but again with a similar increase of pressure losses. These design studies finally led to the choice of wires wrapped around the fuel rods as spacers, which are cheaper, are less detrimental for pressure losses but cause similar coolant mixing, as will be explained later in Chapters 2.3 and 2.6.

The axial distance of grid spacers was optimized exemplarily by Behnke et al. [23] for a test fuel bundle with 4 rods at typical operating conditions of the HPLWR. A cladding temperature difference ΔT between left and right side of a fuel rod will cause bending of the rod towards the hotter side. The maximum thermal deflection b between 2 spacers can be estimated from the axial spacer distance a, the thermal expansion coefficient β of the cladding material and the outer rod diameter d as

$$b = \frac{a^2 \beta \Delta T}{8d}$$

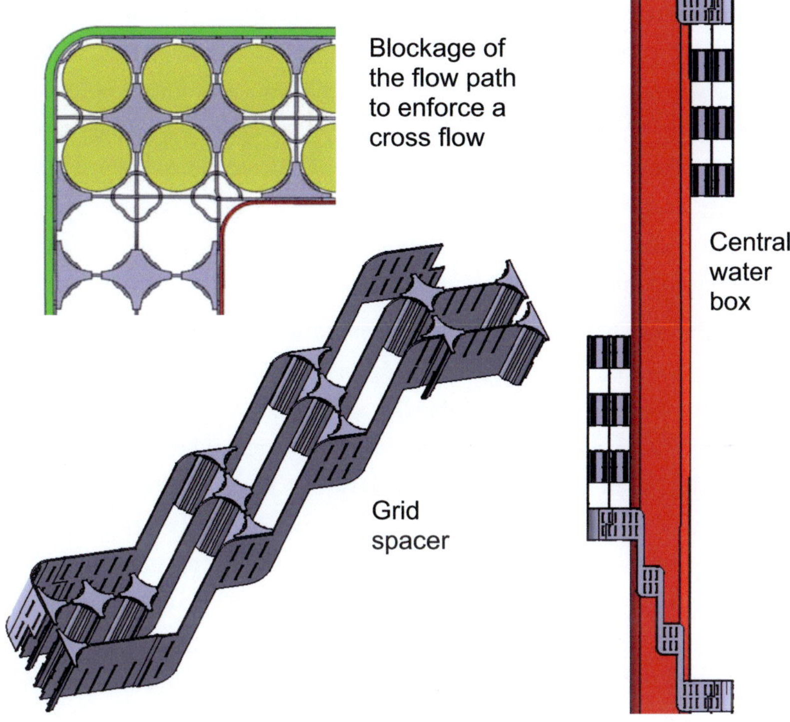

Fig. 1.15: Design of a staircase grid spacer enhancing heat transfer and coolant mixing in the fuel assembly [[20], [21]]

This fuel rod bending, in turn, will narrow the coolant flow path on the hotter side, which will reduce the coolant velocity there and thus increase the cladding temperature further on. Such kind of thermal instability could even continue until the fuel rods come into contact, causing a local burn out there. Behnke [23] recommend, therefore, to limit the maximum thermal deflection b to ¼ of the gap between fuel rods to avoid this effect. The consequences of this criterion on the axial spacer distance a is shown exemplarily in Fig. 1.16, assuming an outer diameter of 8 mm of the stainless steel claddings and a gap of 1.2 mm between them. A temperature difference of 50°C between the side facing a box wall and the central side of a fuel rod, as predicted by Waata [18] would thus require a spacer distance of about 15 cm only. Even though this estimation is somewhat conservative, this example underlines the necessity of excellent coolant mixing inside the assembly, and it indicates that axial spacer distances will need to be significantly smaller in an HPLWR than in pressurized or boiling water reactors.

As a sliding pressure had been envisaged in the HPLWR project, Behnke et al. [[22], [23]] studied the cladding temperatures to be expected at sub-critical pressure. As the coolant will enter the core as sub-cooled liquid but will leave it afterwards as superheated steam, a dryout will occur somewhere in the core unless steam separators are foreseen like in the HDR. Therefore, Behnke et al. included a model for dryout and post-dryout heat transfer at near critical pressure in the sub-channel code MATRA and applied the code exemplarily to a four rod test assembly design. Due to a void drift in the assembly, caused by capillary forces at high steam quality, they predict that the central sub-channel will have a higher void fraction than those ones close to the box wall, which causes the central sub-channel to dry out first.

The average cladding temperature predicted for a test case simulating 60% of the nominal reactor power is shown in Fig. 1.17. A pressure of 15 MPa and a mass flux of 600 kg/m^{2}s have been assumed for a 4 rod test fuel bundle, while the heat flux has been varied between 550 and 750 kW/m^2. As soon as dryout occurs in the central sub-channel 3, the average cladding temperature increase in a first step, shown in Fig. 1.17. Dryout in the wall sub-channels 2 is causing the second step of the average cladding temperature, and the final step is due to dryout in the corner sub-channel 1. While cladding temperatures up to 700°C might be acceptable for some materials, the temperature difference between the central and the corner sub-channel is certainly not. According to Fig. 1.16, a temperature difference of more than 200°C between central and corner sub-channel will bring the fuel rods into contact and a burn out will be unavoidable. The consequence of these studies was finally that the sliding pressure operation was given up.

Outside Europe, there were 2 alternative concepts of fuel assembly design, which do not require any water rods to simplify the design and which should be mentioned here:

Buongiorno [24] proposed a hexagonal fuel assembly with 19 fuel rods of 9.5 mm outer diameter and a pitch of 10.5 mm inside an assembly box with 51.9 mm width (flat to flat). A gap of 16 mm, filled with feedwater, was left between the assembly boxes for moderation, but no additional water rod was foreseen inside the assembly box. Control rods were assumed to run inside these gaps, inserted from the core bottom. A wire wrapped around each fuel rod with an axial pitch of 190 mm was proposed as spacer providing good coolant mixing inside the assembly. However, a 2D neutronic analysis with MCNP of 1/12 of the assembly, assuming a uniform enrichment of 5% for all fuel rods and reflecting boundary conditions, resulted in a rather non-uniform power distribution: the central fuel rod had only 72.6% of the average power, whereas the fuel rods next to the box walls had 117% and 106% of the average power. A refined enrichment and burn-up concept will be needed to optimize this concept.

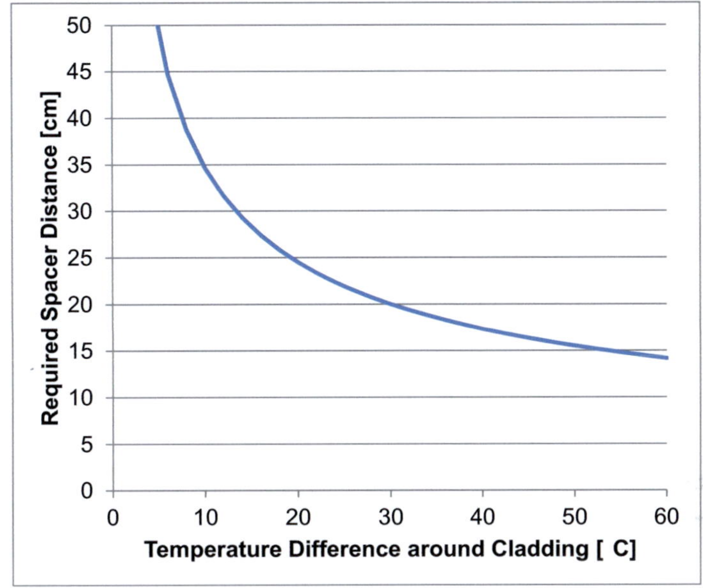

Fig. 1.16: Axial spacer difference required to avoid thermal instabilities.

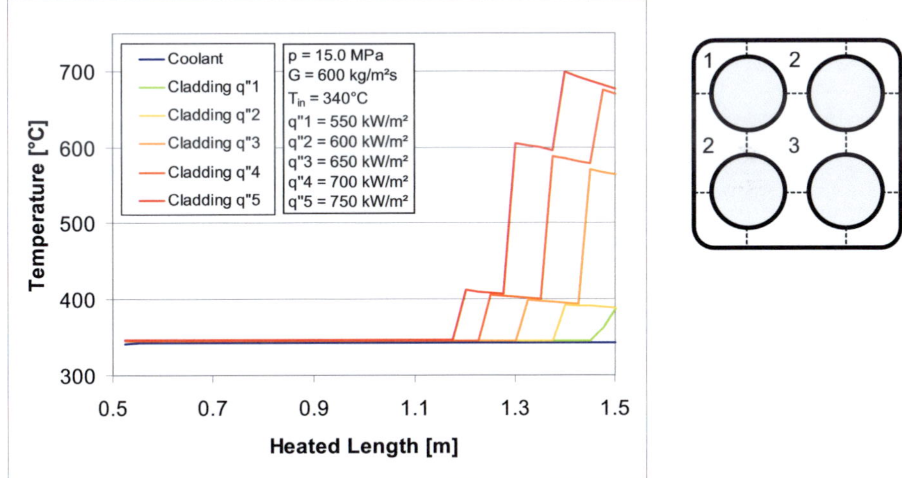

Fig. 1.17: Average cladding temperatures predicted for a 4 rod fuel bundle test at sub-critical pressure of 15 MPa with a mass flux of 600 kg/m^{2}s at 5 different heat flux assumptions [22]

Joo et al. [25] proposed a square fuel assembly with fuel rods of 9.5 mm outer diameter and a pitch of 11.5 mm. Besides the fuel rods, the assembly array of 21x21 rods was filled with 25 cruciform ZrH moderator rods replacing 5 fuel rods each, 24 control rods running inside the cruciform moderator rods, and 28 gadolinia rods as burnable poison. A gap of 10 mm was left between the assembly boxes providing additional moderation. Solid ZrH as an alternative moderator has been proposed repeatedly, but the hydrogen release during accidental conditions must still be seen as an issue, as discussed by Buongiorno and MacDonald [[26], [27]].

1.2.3 Core design concepts

Neither the fissile power nor the coolant mass flow rate can be distributed uniformly over the core cross section. There are a number of radial and axial peaking factors, effecting either the power of individual fuel assemblies or fuel rods or effecting the coolant mass flow rate through assemblies or sub-channels. Schulenberg et al. [28] discuss reasonable target data for these peaking factors and their key parameters, which should be considered for a robust core design, as summarized in Tab. 1.3.

Hot channel factor	axial	radial	Key parameters
Form factors for power profiles			Fuel enrichment and distribution, water density distribution, reflector design and properties, fuel and control rod pattern, burn-up, burnable poisons, …etc
• Assembly peaking factor		1.25	
• Local peaking factor inside assemblies		1.15	
• Axial power factor	1.6		
• Uncertainties		1.2	Material properties of coolant and claddings, physical modeling, hydraulic modeling, heat transfer coefficient, geometry tolerances
Allowances		1.15	Power control, flow control, pressure control, inlet temperature control
Total	1.6	2.0	

Tab. 1.3: Target hot channel factors for a robust core design [28]

While the axial form factor is responsible for fuel and cladding temperatures in the core, the radial form factor needs to be multiplied with the enthalpy rise in the core. Firstly, an assembly peaking factor is needed to account for differences in power distribution of assembly clusters and coolant mass flux distributions in the core. Reasons for these differences are the fuel composition and distribution, water density distribution, size and distribution of sub-channels, neutron leakage and reflector effects, burn-up effects, effect of control rod positioning or effects due to the use of burnable poisons. Secondly, a local peaking factor is needed to account for similar differences inside assemblies, which has to be multiplied with the assembly form factor to yield the hottest sub-channel in the hottest assembly. An uncertainty factor including all statistical uncertainties of the core design accounts for material uncertainties, fluid properties, uncertainties of the neutron physical modeling, heat transfer uncertainties, uncertainties of the thermal-hydraulic modeling, scattering of the inlet temperature distribution, manufacturing tolerances, deformations during operation, or measurement uncertainties of the installed measurement systems. Such uncertainties do not add linearly but, as most of them are independent statistical parameters, rather as a sum of variances. Finally, a further hot channel factor is needed to account for allowances for plant operation, such as power control, flow control, pressure control, or inlet temperature control leading to small but allowable transients during operation.

If all these four factors (excluding the axial form factor) are multiplied, we get a total hot channel factor of 2.0 which differs the enthalpy rise of the hottest sub-channel under worst case conditions from the nominal sub-channel. Due to the particular fluid properties of supercritical water, however, these hot channel factors have a surprising effect on the peak coolant temperature. To illustrate this, we plot in Fig. 1.18 the temperature of supercritical water at 25 MPa vs. its enthalpy. The plateau at 384°C characterizes the transition from liquid like to steam like conditions with a peak in the specific heat. The average coolant heat up from 280°C to 500°C, listed as target data in Tab. 1.2, corresponds with a coolant enthalpy rise from 1230 kJ/kg to 3166 kJ/kg which yields an enthalpy difference of 1936 kJ/kg. If we multiply this enthalpy difference by a hot channel factor of 2, we get a local peak coolant outlet enthalpy of 5102 kJ/kg, which corresponds with a local peak coolant outlet temperature of around 1200°C. This would be far beyond any reasonable material limit.

Different solutions to this issue have been discussed by Schulenberg and Starflinger [29]. A simple and straight forward method could be to limit the average coolant heat up from 280°C to 380°C only. This corresponds with an enthalpy rise 1230 kJ/kg to 1936 kJ/kg, resulting in a local peak coolant outlet temperature of 405°C if the hot channel factor of 2 is applied again. Obviously, we take advantage now of the temperature plateau at 384°C such that any enthalpy rise in a local hot channel is flattened effectively. The method is similar to a PWR where the average coolant outlet temperature is also chosen to be just below the boiling temperature, such that a peak coolant temperature will be flattened by local boiling. As a drawback, however, the hot coolant at core outlet is still liquid like in this supercritical water concept and a closed loop primary system with steam generators will be required to provide steam for the turbines.

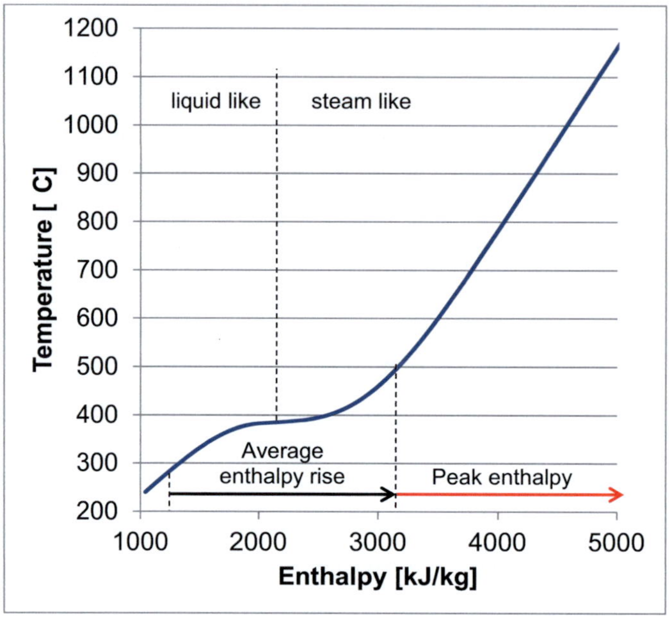

Fig. 1.18: Temperature vs. enthalpy of supercritical water at 25 MPa

A core design concept following this idea has been worked out by Vogt et al. [30], called the PWR-SC. The reactor core was designed for a thermal power of 2000 MW with a reactor inlet temperature of 280°C and an outlet temperature of 380°C. The assembly design sq2.1, Fig. 1.10, was used and combined to an assembly cluster of 9 assemblies each with common head and foot piece to ease handling during fuel shuffling. Details of the foot piece are shown in Fig. 1.19 (left). Moderator water inside the water rods was supplied from the top of the core and was running downwards, to be released inside the foot piece and mixed there with feedwater supplied through the downcomer of the reactor. Hofmeister et al. [31] analyzed the flow structure in the mixing chamber of the foot piece with CFD to predict the coolant temperature profile at the core inlet: the residual temperature non-uniformity was less than 3°C. Details of the head piece are shown in Fig. 1.19 (right). The head piece was extended with a cylinder with wide openings to release the produced hot coolant to a plenum which could be mounted over the head pieces. The water rods were extended to the top of the cylindrical extension to guide the moderator water through the hot plenum without leakage. C-Rings were sealing the hot plenum from the feedwater as explained by Hofmeister et al.

[15]. Control rods were inserted from the top into 5 of these 9 water rods through the square openings shown in Fig. 1.19 (right).

The proposed reactor design is shown in Fig. 1.20. Design of the core barrel, of control rod drives and of control rod guide tubes could be taken from PWR design, whereas the reactor pressure vessel needed to be designed with thicker walls to account for the higher pressure of 25 MPa. Details of the pressure vessel design and its structural analysis have been published by Fischer et al. [32]. With an inner diameter of 3.38 m, a wall thickness of 319 mm has been required in the lower cylindrical part and of 510 mm in the upper part around the inlet and outlet flanges. The co-axial feedwater supply around the hot outlet line, as indicated in Fig. 1.8, has been replaced by a slow purge flow around the hot inner tube and 4 separate inlet flanges instead to minimize heat losses of the hot coolant. The hot plenum mounted over the head pieces, shown in green color in Fig. 1.20, could be removed after extracting the hot tubes radially out of the plenum. The foot pieces were standing on a core support plate. Removable orifices underneath the foot piece were adjusting the coolant mass flow to the individual power of each assembly cluster. This design was actually intended to be used also for higher coolant outlet temperatures and could eventually be simplified for the low temperature application considered here.

A coupled, steady state analysis of power and coolant temperature distribution has been performed by Vogt [33] with the neutronic code MCNP5 and the sub-channel code STAFAS [14] assuming fresh fuel with an enrichment between 3.75% in the centre up to 5.5% in the outer core regions. The core power density was approx. 100 MW/m^3. The local core outlet temperature was reaching up to 395°C in selected assemblies, not including yet uncertainties and allowances as listed in Tab. 1.3. If these are added as proposed in Tab. 1.3, a radial hot channel of 2.05 was predicted, resulting in a peak coolant temperature of 416°C.

The primary loop was intended to produce superheated steam of 370°C at 7.5 MPa in the steam generators of the secondary system, resulting in a net power of 748.8 MW, which yields a net efficiency of 37.5% or an improvement of 2% pts. compared with a conventional PWR at same condenser pressure. The required power of the 4 primary pumps would be reduced to ¼ of the power of a comparable PWR, thanks to the significant increase of the enthalpy rise in the core. The lower mass flow rate in the primary system would even reduce the mass of all pressure tubes outside the reactor to less than 50% despite the higher pressure. The steam turbines would require 20% less mass flow rate at a given net power of the power plant which would allow descaling of all components of the secondary steam cycle. The steam dryer between high pressure and low pressure turbines could even be omitted thanks to the higher temperatures. In total, Vogt et al. [30] expect significant cost reductions for this concept, but the design has not been sufficiently detailed up to now to quantify this precisely.

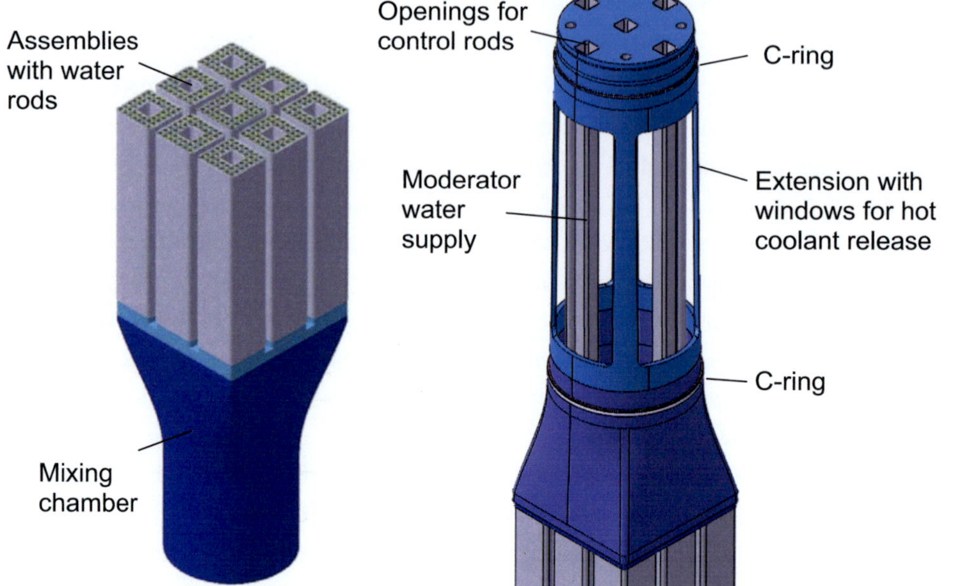

Fig. 1.19: Foot piece (left) and head piece (right) of the PWR-SC fuel assembly cluster [30]

If higher coolant outlet temperatures shall be achieved, a heat up process in several steps becomes necessary, with coolant mixing between each step. Fig. 1.21 illustrates different core design concepts in principle. The concept of the PWR-SC is comparable, in principle, with the single pass core shown on the left. The coolant enters the core from the bottom and leaves it at the top like in a PWR.

The two pass core is heating up the coolant in a first step in fuel assemblies in the outer core region with a downward flow. It is mixed then underneath the core and finally heated up to the core outlet temperature in assemblies of the central core region. Such a flow path has been chosen by Yamaji et al. [6] for their SCWR design. Cans around the head pieces of assembly clusters in the steam plenum above the core must be provided to allow the feedwater to penetrate it. Such cans have been designed by Schulenberg et al. [34] for the reactor design shown in Fig. 1.19 and 1.20, demonstrating that the core design can easily be modified to enable such a flow path. Lower peak coolant temperatures or higher core outlet temperatures can be achieved with a three pass core, sketched on the right hand side of Fig. 1.21. Here, the coolant enters the core first from below, rises upwards in the central fuel assemblies of the core, and is mixed then in a steam plenum above the core. A second heat-up step is provided in a downward flow in fuel assemblies surrounding them, mixed again in a second, annular mixing plenum underneath the core, and finally heated up to the core outlet temperature with an upward flow at the core periphery.

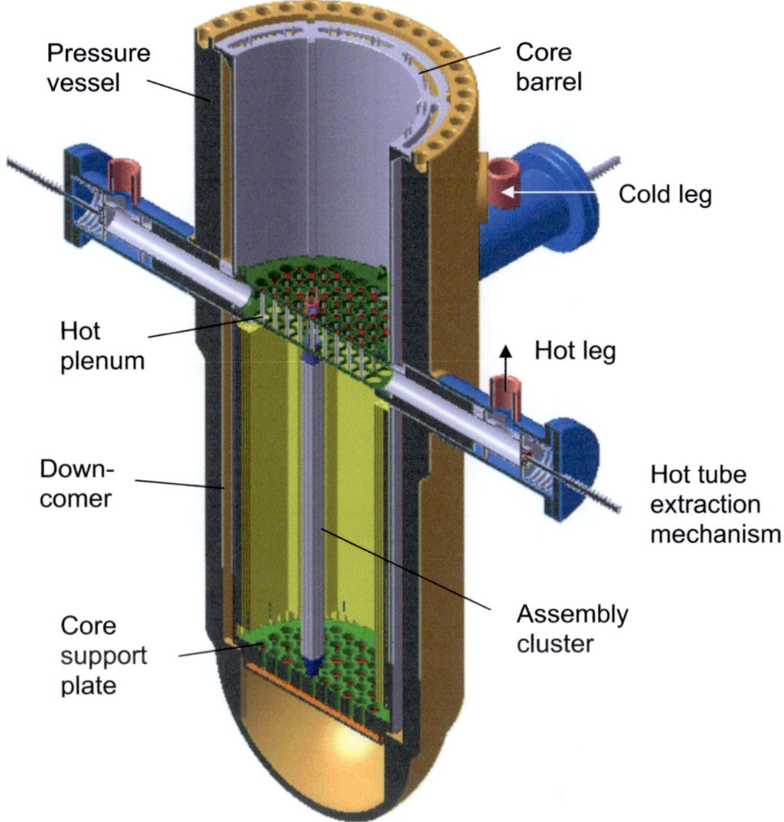

Fig. 1.20: Design of the PWR-SC: reactor pressure vessel and core design [32]

The advantage of such multistep heat-up is illustrated in Fig. 1.22. The average enthalpy rise is shown with blue arrows and the additional heat up in the hot channel with red arrows. Mixing after each heat-up step eliminates the hot streak and the next heat up step is starting again from a homogenous mixture. Using the target hot channel factor of 2 given in Tab. 1.3, Fig. 1.22 shows that an average core outlet temperature of 500°C could be achieved without exceeding peak coolant temperatures of 600°C in any heat-up step. The method of multiple heat-up steps is not new, though, but a standard procedure in fossil fired boiler design. We will show in Chapter 2 how such a core can be designed and discuss if the envisaged hot channel factors can be realized.

These generic design studies were the basis of a second phase of the HPLWR project, aiming still for the design target set in Tab. 1.2, as will be presented next.

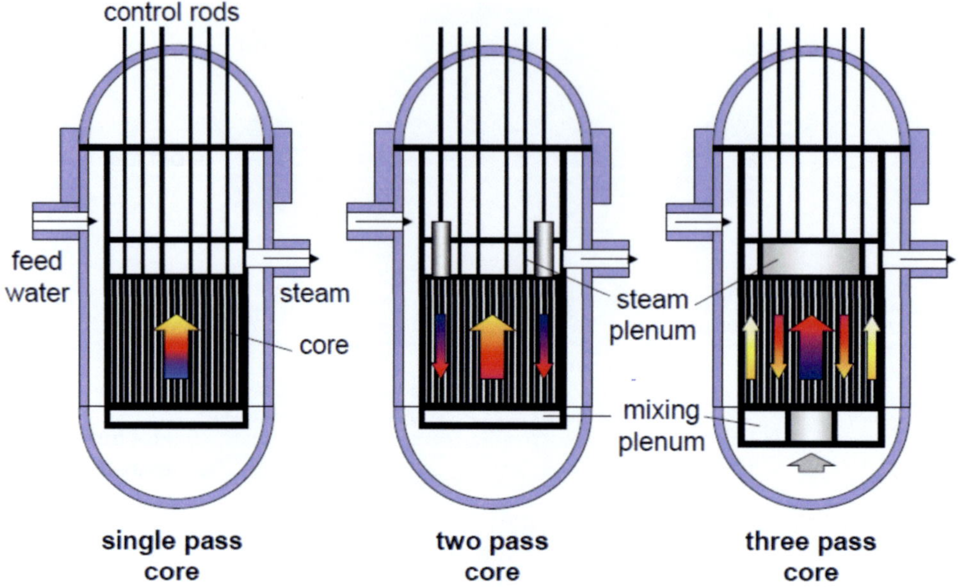

Fig. 1.21: Sketch of different core design concepts for higher coolant heat up [29]

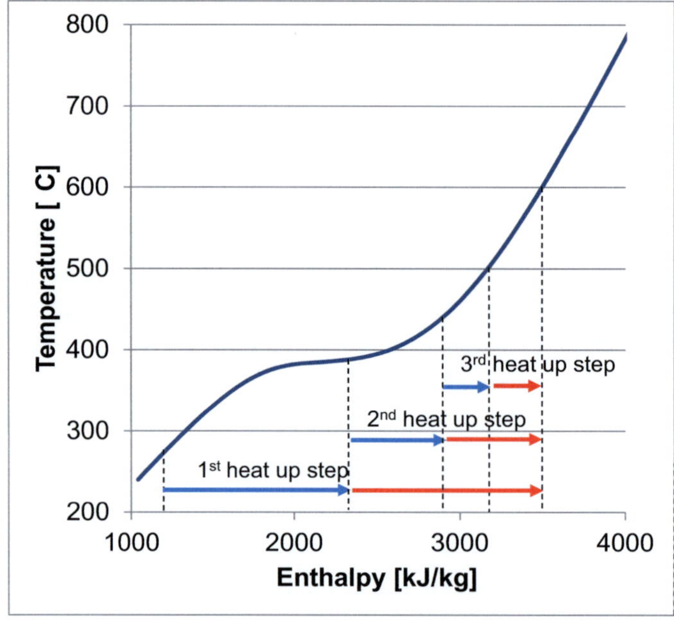

Fig. 1.22: Heat up in three steps to 500°C average outlet temperature with 600°C peak coolant temperature; in blue: average enthalpy rise; in red: additional peak enthalpy rise

1.2.4   Phase 2 of the HPLWR project

Since the supercritical water cooled reactor (SCWR) became one of the six nuclear energy systems of the Generation IV International Forum, the worldwide resources developing SCWR concepts were enlarged, research organizations and industry from USA, Canada, Japan, South Korea, Europe and finally China and Russia became interested to work jointly on SCWR subjects, and a system arrangement for international research and development of SCWR nuclear systems was signed by Canada, Japan and Euratom in 2006. With the new perspective of long term collaboration, the European consortium of the HPLWR project sketched a roadmap for HPLWR development, published by Starflinger et al. [35] in 2003. Within this frame, a second phase of the HPLWR project was planned, described by Schulenberg et al. [36], in which the HPLWR reactor and power plant should be designed in more detail to assess the future potential, based on ideas and concepts of the first phase. The project was accompanied by a technology program for materials and thermal-hydraulics of supercritical water and by an education and training program for this innovative reactor technology. The European project was finally launched in Sept. 2006 for a duration of 3.5 years, supported by the European Commission in their 6th Framework Programme.

Structure and work plan of this new project "HPLWR Phase 2" were outlined by Starflinger et al. [37]. The design work was structured into three major phases of 1 year each. The first design proposal was to be reviewed and frozen after a mid-term assessment, to be analyzed in the second year, and the third year was left then for a second design iteration including repeated analyses. Of course, the status reached by the end of the project cannot considered to be final then, but it was expected to be detailed enough to assess the concept with respect to the criteria defined by the Generation IV International Forum [38]. Design tasks included core design, reactor design, safety systems and steam cycle components, and analyses covered neutronic analyes, thermal-hydraulic analyses, structural analyses and system analyses for steady state and transient conditions. The technology program was focused on heat transfer of supercritical water in the reactor core, on materials for fuel claddings, and on the water chemistry needed.

By the mid-term assessment after 1 year, summarized by Starflinger et al. [39], the first coupled neutronic/thermal-hydraulic analyses of the core were available for full load, steady state conditions. They showed that the envisaged power profile and coolant density distribution were feasible. CFD analyses of coolant mixing inside assemblies as well as in the mixing chambers above and below the core predicted an acceptable temperature distribution at the inlet of each heat up step. Stress and deformation analyses of the reactor pressure vessel, the major reactor internals and of the assembly boxes indicated areas for design optimization which were to be fixed with the next design iteration.

As part of the technology program, the physics of deterioration of heat transfer of a flow of supercritical water with low mass flux through a tube with high heat flux was studied with CFD: If the boundary layer was well resolved, and if physical properties of supercritical water were included properly in the analysis, the numerical simulation could model the observed phenomena at least qualitatively. A numerical study of turbulence enhancement by ribs on the heated wall indicated this measure as appropriate to avoid the deterioration of heat transfer.

A first design proposal of a containment for the High Performance Light Water Reactor had been worked out within the first year and first transient analyses of design basis accidents were started.

The following chapters shall summarize the results achieved by the end of this project in Feb. 2010 to document all design work which had been performed until then. The final assessment, concluded by the end of the project, is summarized in the final report [40]. It was presented and discussed with other partners of the Generation IV International Forum in the 5th International Symposium on SCWR in March 2011, published by Starflinger et al. [41], Schulenberg et al. [42], and Andreani et al. [43], supported by detailed reports of other HPLWR partners. The enormous number of publications about HPLWR design details and their analyses, however, made it difficult to keep an overview, which gave reason to summarize them in a comprehensive form in this book.

1.3   References

[1]   Y. Oka, Review of high temperature water and steam cooled reactor concepts, SCR-2000, Paper 104, Tokyo, Nov. 6-8, 2000

[2]   R. Dörfler, Aufbau der Gesamtanlage HDR, Atomwirtschaft XIV/11 , pp. 534-537, Nov. 1969

[3]   K. Traube, L. Seyfferth, Der Heissdampfreaktor - Konstruktion und Besonderheiten, Atomwirtschaft XIV/11, pp. 539-542, Nov. 1969

[4]   J. Höchel, W. Fricke, Besonderheiten bei der Entwicklung und Fertigung der HDR-Brennelemente, Atomwirtschaft XIV/11, pp. 549-555, Nov. 1969

[5] K. Dobashi, Y. Oka, S. Koshizuka, Core and plant design of the power reactor cooled and moderated by supercritical water with single tube water rods, Ann. Nucl. Energy, Vol. 24, 16, pp. 1281-1300, 1997

[6] A. Yamaji, Y. Oka, S. Koshizuka, Three dimensional core design of high temperature supercritical-pressure light water reactor with neutronic and thermal-hydraulic coupling, Journal of Nuclear Science and Technology, 42, 1, pp. 8-19 (2005).

[7] L. Cao, Y. Oka, Y, Ishiwatati, Z. Shang, Fuel, core design and subchannel analysis of a Super Fast Reactor, Journal of Nuclear Science and Technology, 45, 2, pp. 1-11 (2008)

[8] Y. Oka, S. Koshizuka, Y. Ishiwatari, A. Yamaji, Super Light Water Reactors and Super Fast Reactors, Springer, ISBN 978-1-4419-6034-4, 2010

[9] D. Squarer, T. Schulenberg, D. Struwe, Y. Oka, D. Bittermann, N. Aksan, C. Maraczy, R. Kyrki-Rajamäki, A. Souryi, P. Dumaz, High performance light water reactor, Nuclear Engineering and Design 221, pp. 167-180 (2003)

[10] D. Bittermann, D. Squarer, T. Schulenberg, Y. Oka, P. Dumaz, R. Kyrki-Rajamäki, N. Aksan, C. Maraczy, Y. Souyri, Potential plant characteristics of a high performance light water reactor, Proc. ICAPP 03, Paper 3046, Cordoba, Spain, May 4-7, 2003

[11] K. Ehrlich, J. Konys, L. Heikinheimo, Materials for High Performance Light Water Reactors, Journal of Nuclear Materials 327 (2004) 140-147

[12] G. Rimpault, C. Maraczy, R. Kyrki-Rajamäki, Y. Oka, T. Schulenberg, Core design feature studies and research needs for high performance light water reactors, Proc. ICAPP 03, Paper 3194, Cordoba, Spain, May 4-7, 2003

[13] O. Antoni, P. Dumaz, Preliminary calculations of supercritical light water reactor concept using the CATHARE code, Proc. ICAPP 03, Paper 3146, Cordoba, Spain, May 4-7, 2003

[14] X. Cheng, T. Schulenberg, D. Bittermann, P. Rau, Design analysis of core assemblies for supercritical pressure conditions, Nuclear Engineering and Design 223, pp. 279-294 (2003)

[15] J. Hofmeister, T. Schulenberg, J. Starflinger, Optimization of a fuel assembly for a HPLWR, Proc. ICAPP 05, Paper 5077, Seoul, Korea, May 15-19, 2005

[16] J. Hofmeister, Auslegung eines Brennelements für einen Leichtwasserreaktor mit überkritischen Dampfzuständen, Dissertation, University of Stuttgart, Forschungszentrum Karlsruhe FZKA 7248, 2006

[17] C. Waata, T. Schulenberg, X. Cheng, E. Laurien, Coupling of MCNP with a sub-channel code for analysis of a HPLWR fuel assembly, Proc. NURETH-11, Paper 021, Avignon, France, Oct. 2-6, 2005

[18] C. Waata, Coupled neutronics / thermal-hydraulics analysis of a high performance light water reactor fuel assembly, Dissertation, University of Stuttgart, Forschungszentrum Karlsruhe FZKA 7233, 2006

[19] J. Hofmeister, C. Waata, J. Starflinger, T. Schulenberg, E. Laurien, Fuel assembly design study for a reactor with supercritical water, Nuclear Engineering and Design 237, pp. 1513-1521, 2007

[20] A. Bastron, J. Hofmeister, L. Meyer, T. Schulenberg, Enhancement of heat transfer in HPLWR fuel assemblies, Proc. GLOBAL 2005, Paper 036, Tsukuba, Japan, Oct. 9-13, 2005

[21] L. Meyer, A. Bastron, J. Hofmeister, T. Schulenberg, Enhancement of heat transfer in fuel assemblies of high performance light water reactors, Journal of Nuclear Science and Technology, Vol. 44, No. 3, pp. 1-7 (2007)

[22] L. Behnke, S. Himmel, C. Waata, E. Laurien, T. Schulenberg, Prediction of heat transfer for a supercritical water test with a four pin fuel bundle, Proc. ICAPP 06, Paper 6211, Reno, USA, June 4-8, 2006

[23] L. Behnke, Vorhersage kritischer Siedezustände in Brennelementen von Leichtwasser-reaktoren bei variablem Druck, Dissertation University of Stuttgart, 2009

[24] J. Buongiorno, An alternative SCWR design based on vertical power channels and hexagonal fuel assemblies, Proc. GLOBAL 03, pp. 1155-1162, New Orleans, USA, Nov. 17-20, 2003

[25] H.K. Joo, K.M. Bae, H.C.Lee, J.M. Noh, Y.Y. Bae, A conceptual core design with a rectangular fuel assembly for a thermal SCWR system, Proc. ICAPP 05, Paper 5223, Seoul, Korea, May 15-19, 2005

[26] J. Buongiorno, P. MacDonald, Study of solid moderators for the thermal-spectrum supercritical water-cooled reactor (neutronics), Proc. ICAPP 03, Paper 3328, Cordoba, Spain, May 4-7, 2003

[27] J. Buongiorno, P. MacDonald, Study of solid moderators for the thermal-spectrum supercritical water-cooled reactor (thermal-mechanics and costs), Proc. ICAPP 03, Paper 3329, Cordoba, Spain, May 4-7, 2003

[28] T. Schulenberg, J. Starflinger, J. Heinecke, Three pass core design proposal for a high performance light water reactor, Progress in Nuclear Energy 50, pp. 526-531 (2008)

[29] T. Schulenberg, J. Starflinger, Core design concepts for high performance light water reactors, Nuclear Engineering and Technology, Vol. 39, 4, pp. 1-8 (2007)

[30] B. Vogt, K. Fischer, J. Starflinger, E. Laurien, T. Schulenberg, Concept of a pressurized water cooled reactor cooled with supercritical water in the primary loop, Nuclear Engineering and Design 240, pp. 2789-2799 (2010)

[31] J. Hofmeister, E. Laurien, A. Class, T. Schulenberg, Turbulent mixing in the foot piece of a HPLWR fuel assembly, Proc. GLOBAL 2005, Paper 066, Tsukuba, Japan, Oct. 9-13, 2005

[32] K. Fischer, J. Starflinger, T. Schulenberg, Conceptual design of a reactor pressure vessel and its internals for a HPLWR, Proc. ICAPP 06, Paper 6098, Reno, USA, June 4-8, 2006

[33] B. Vogt, Analyse eines Druckwasserreaktors mit überkritischem Wasser als Kühlmittel, Dissertation, University of Stuttgart, Forschungszentrum Karlsruhe FZKA 7428, 2008.

[34] T. Schulenberg, K. Fischer, J. Starflinger, Review of design studies for high performance light water reactors, 3rd Int. Symposium on SCWR, Paper P024, Shanghai, China, March 12-15, 2007

[35] J. Starflinger, N. Aksan, D. Bittermann, L. Heikinheimo, C. Maraczy, G. Rimpault, T. Schulenberg, Roadmap for supercritical water cooled reactor R&D in Europe, GLOBAL 03, pp. 1137-1142, New Orleans, USA, Nov. 17-20, 2003

[36] T. Schulenberg, J. Starflinger, N. Aksan, D. Bittermann, L. Heikinheimo, Supercritical water reactor research in the GIF context: current status and future prospects with emphasis on European activities, Proc. FISA 2006, Luxembourg, March 13-16, 2006

[37] J. Starflinger, T. Schulenberg, P. Marsault, D. Bittermann, C. Maraczy, E. Laurien, J.A. Lycklama, H. Anglart, N. Aksan, M. Ruzickova, L. Heikinheimo, European research activities within the project „High Performance Light Water Reactor Phase 2", Proc. ICAPP 07, Paper 7146, Nice, France, May 13-18, 2007

[38] A Technology Roadmap for Generation IV Nuclear Energy systems, issued by the US DOE Nuclear Energy Research Advisory Committee and the Generation IV International Forum, GIF-002-00, December 2002

[39] J. Starflinger, T. Schulenberg, P. Marsault, D. Bittermann, C. Maraczy, E. Laurien, J.A. Lycklama, H. Anglart, M. Andreani, M. Ruzickova, L. Heikinheimo, Results of the mid-term assessment of the "High Performance Light Water Reactor Phase 2" project, Proc. ICAPP 09, Paper 9268, Tokyo, Japan, May 10-14, 2009

[40] J. Starflinger, T. Schulenberg, P. Marsault, D. Bittermann, E. Laurien, C. Maraczy, H. Anglart, J.-A. Lycklama, M. Andreani, M. Ruzickova, T. Vanttola, A. Kiss, M. Rohde, R. Novotny, High Performance Light Water Reactor Phase 2, Public Final Report – Assessment of the HPLWR Concept, 6th Framework Programme, Contract No. FI60-036230, 2010, www.hplwr.eu

[41] J. Starflinger, T. Schulenberg, D. Bittermann, M. Andreani, C. Maraczy, Assessment of the High Performance Light Water Reactor concept, 5th Int. Sym. SCWR (ISSCWR-5), P056, Vancouver, Canada, March 13-16, 2011

[42] T. Schulenberg, C. Maraczy, W. Bernnat, J. Starflinger, Assessment of the HPLWR thermal core design, 5th Int. Sym. SCWR (ISSCWR-5), P057, Vancouver, Canada, March 13-16, 2011

[43] M. Andreani, D. Bittermann, P. Marsault, O. Antoni, K. Kereszturi, M. Schlagenhaufer, A. Manera, M. Seppälä, J. Kurki, Evaluation of a preliminary safety concept for the HPLWR, 5th Int. Sym. SCWR (ISSCWR-5), P068, Vancouver, Canada, March 13-16, 2011

2 Core design

A core which is cooled with supercritical water can be designed with a thermal or with a fast neutron spectrum, in general. The option of a thermal spectrum requires additional water as moderator, because of the low density of superheated steam, which can be provided in water rods inside fuel assemblies or in gaps between assembly boxes. Examples can be seen at latest boiling water reactor design or in the Super Light Water Reactor concept by Oka et al. [1], using supercritical water. If these gaps and water rods are omitted, the neutron spectrum will become fast which simplifies the design and increases the core power density. A general safety concern of the fast core option, however, is the reactivity increase if the core should be voided under accidental conditions. Such a reactivity increase must definitely be avoided by suitable core design, for which the addition of some solid moderator, an increased neutron leakage and a heterogeneous arrangement of seed and blanket assemblies are common measures. Oka et al. [1] give an example with their Super Fast Reactor concept. The HPLWR core design described here shall provide a thermal neutron spectrum in the entire core.

2.1 Design target

Aiming at a net electric power of around 1000MW and a net efficiency of almost 44%, the target thermal power of the reactor core needs to be 2300MW, confirmed by steam cycle analyses of Brandauer et al. [2]. Early cycle studies by Dobashi et al. [3] indicated an optimum thermal efficiency at a feedwater temperature of 280°C which was kept also for the present study. The target core outlet temperature was chosen as 500°C which is still rather low for a once through steam cycle with single reheat, compared with latest fossil fired power plants, but appears to be challenging enough with regard to available fuel cladding materials. Their peak temperature limit was targeted at 630°C which is not only a challenge for oxidation and corrosion protection, but also for their creep strength and resistance to stress corrosion cracking. The fuel centerline temperature is a function of the linear power of the fuel rod. The latter one has been limited to 39kW/m under nominal conditions. To be competitive with respect to latest pressurized water reactors, the target burn up should be at least 60 MWd/t_{HM}. Like with boiling water reactors, boron acid cannot be used to compensate the excess reactivity at the beginning of a burn-up cycle, so that burnable absorbers like Gd must be used instead. The target power and temperatures result in a coolant mass flow rate of 1179kg/s. Schlagenhaufer et al. [4] suggest a feedwater pressure of 25MPa for all load conditions which keeps some margin from the critical pressure of 22.1MPa.

2.2 General design strategy

These target data differ from conventional light water reactors not only by the higher pressure and by the core outlet temperature, but also by a significantly higher enthalpy rise in the core. Indeed, the difference between life steam enthalpy and feedwater enthalpy of 1936kJ/kg exceeds the one of pressurized water reactors by around a factor of 8. Assuming an overall hot channel factor of 2 between the peak and the average coolant heat-up, as discussed in chapter 1.2.3, this enthalpy rise would result in peak coolant temperatures of 1200°C which is far beyond the target temperature limit. A strategy to overcome this issue can be learned from fossil fired boiler design. These boilers are characterized by multiple heat-up steps with intensive coolant mixing between them to eliminate hot streaks.

Schulenberg et al. [5] applied such a strategy for a thermal core layout with a first heat-up of the coolant as moderator water, comparable with the economizer of a fossil fired boiler. The second heat-up should be in the evaporator assemblies in the center of the core, followed by coolant mixing in a plenum above the core. From there, the coolant is directed downwards in assemblies of the first superheater, surrounding the evaporator, to be mixed again in an annular chamber underneath the core. Final heat-up to the envisaged core outlet temperature of 500°C was proposed to happen in a second superheater stage with upward flow again in assemblies at the core periphery. Assuming a hot channel factor of 2 for each heat-up step, as an initial guess, the power ratio of evaporator to superheater 1 to superheater 2 should be around 4:2:1 to reach the same peak coolant temperature in each region. The proposed core layout is trying to reach this power ratio by placing the second superheater at the core periphery where the neutron leakage is reducing the neutron flux anyway. The concept has been worked out to a substantial detail which will be discussed below.

Starting point of the thermal core design has been the assembly design proposal Sq 2.1 of Hofmeister et al. [6], Fig. 1.10, with 40 fuel rods of 8mm outer diameter and a single water box replacing 9 fuel rods, which are housed in an assembly box and grouped to a cluster of 9 assemblies with common head and foot piece to ease handling. Control rods are running from the top into 5 of the 9 water boxes of a cluster.

The core design concept assumes that 50% of the coolant supplied through 4 flanges to the reactor pressure vessel (RPV) is taken first as moderator water to run downwards through these water boxes, to be released through the foot pieces of the assembly clusters to the gap volume between the assembly boxes. There, it rises upwards to serve again as moderator water outside the assembly boxes. It is collected at the top of the core to cool the radial core reflector with a downward flow, before it is mixed with the remaining 50% of the coolant in the core inlet chamber underneath the core. The following three heat-up steps comprise an evaporator region formed by 52 assembly clusters in the core centre, where the coolant changes its density from liquid like to steam like conditions, followed by an upper mixing chamber above the core. Another 52 assembly clusters with downward flow surround the evaporator region and serve as the first superheater. After a second mixing in a lower mixing chamber underneath the core, the coolant is finally heated up to 500°C in a second superheater region formed by 52 assembly clusters at the core periphery. The coolant flow path is sketched in Fig. 2.1. Blue arrows indicate the feedwater which is used as moderator water before entering the core. Red arrows indicate the coolant running inside the assemblies and through the mixing chambers. The arrangement of evaporator and superheater clusters in the core is shown with different colours in Fig. 2.2. Circles are indicating the head pieces of clusters with 9 assemblies each. Some key design data of this concept are listed in Tab. 2.1.

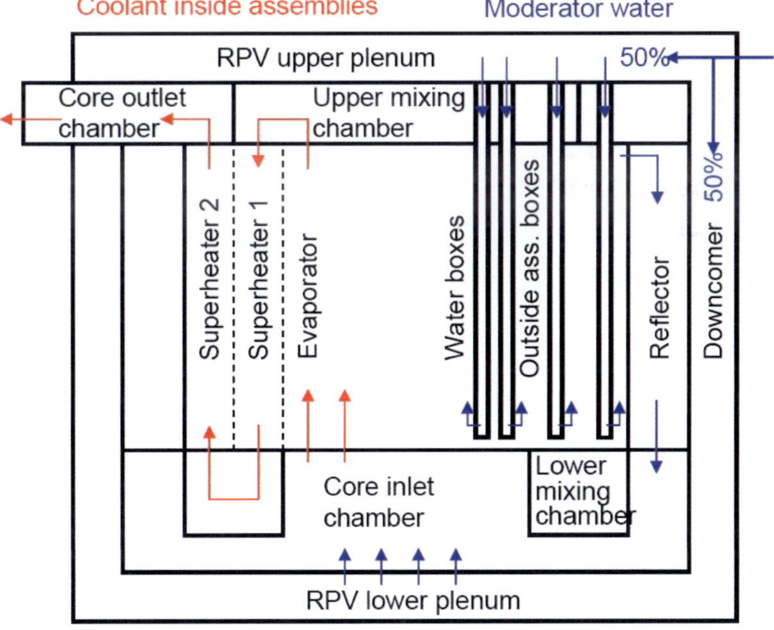

Fig. 2.1: Sketch of the coolant flow path in the thermal core design [51]

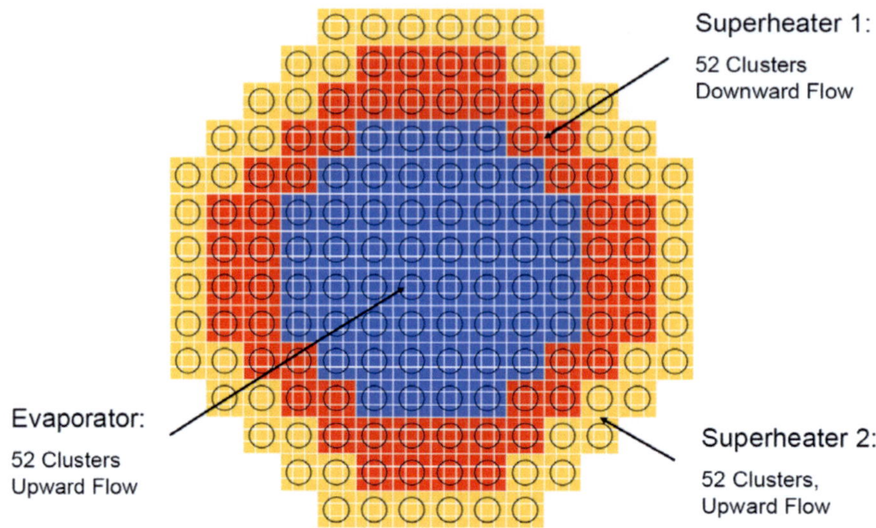

Fig. 2.2: Arrangement of evaporator and assembly clusters in the thermal core

Fuel rods per assembly	40
Water boxes per assembly	1
Assemblies per cluster	9
Evaporator assembly clusters	52
Superheater 1 assembly clusters	52
Superheater 2 assembly clusters	52
Fuel cladding outer diameter	8 mm
Cladding thickness	0.5 mm
Fuel rod pitch / diameter ratio	1.18
Control rods per cluster	5
Active core height	4200 mm
Total core height	5331 mm
Water box wall thickness	2 mm
Water box outer width	26.9 mm
Assembly box wall thickness	3 mm
Assembly box inner width	67.5 mm
Gap between assembly boxes	9 mm
Moderator mass flow fraction	50 %
Flow direction in water boxes	downwards
Flow direction between assembly boxes	upwards
Flow direction in the radial reflector	downwards
Spacer concept	wire wraps
Wire pitch	200 mm
Wire diameter	1.34 mm
Control rod absorber material	B_4C
Average core power density	57.3 MW/m^3
Average coolant mass flux	1600 kg/m^{2}s

Tab. 2.1: Key data of the HPLWR core design

2.3 Mechanical design

Beyond 390°C, the coolant density is less than 200kg/m^3, hardly enough to produce a thermal neutron spectrum. Therefore, colder feedwater is foreseen as moderator water to run inside moderator boxes in the fuel assemblies and in gaps between assembly boxes. As Schulenberg et al. [5] estimated a pressure drop of about 0.5MPa from core inlet to its outlet, but the mass of structural material in the core should be minimized to limit the neutron absorption, Hofmeister et al. [6] concluded that the fuel assemblies should be small, preferably with 40 fuel pins each and a single moderator box in their centre to enable a small wall thickness of moderator and assembly boxes. To ease handling during maintenance, they recommended to group 9 assemblies to a cluster with common head and foot piece. Fischer et al. [7] adopted this idea to design a fuel assembly cluster for the three pass core as described above, preferably such that clusters can be exchanged between evaporator and superheater positions. Wire wraps were proposed as grid spacers to improve coolant mixing in both flow directions. The clusters can be disassembled at their foot piece to exchange single fuel rods for repair. Control rods shall be inserted from the top of the core. They run inside 5 of the 9 moderator boxes of each cluster.

For illustration, Fig. 2.3 (left) shows cut out view of a single fuel assembly. The assembly box and the water box are made of a stainless steel sandwich construction with an internal honeycomb structure filled with Zirconia to improve the thermal insulation and to reach the envisaged stiffness of less than 0.5mm deflection towards the fuel rods under an outside pressure load of 500kPa, as reported by Herbell and Himmel [8]. Details of the box design are shown in Fig. 2.3, centre and right. A venting hole per honeycomb, open to the colder side, is reducing the pressure load acting on the honeycomb structure. The corner pieces are made of solid stainless steel structures to reduce peak stresses there.

The assembly cluster is shown in Fig. 2.4 with the major part of the 9 assemblies being cut out. The length of the assembly boxes between head and foot piece is 5331mm. The common head piece of a fuel assembly cluster is shown in more detail in Fig. 2.5. The upper mixing chamber is mounted over the window element and sealing rings (C-rings) avoid ingress of moderator water into the mixing chamber. Moderator water enters into the water boxes through orifices in the top of the head piece, Fig. 2.5, left. A common spider for 5 control rods can be coupled with the control rod drive. The inner assembly box is welded with the bottom plate of the head piece. It will carry the weight of the foot piece with all assemblies standing on it when the cluster will be lifted. The other 8 assembly boxes are sliding with their round extensions in the bottom plate using piston rings to seal against ingress of moderator water.

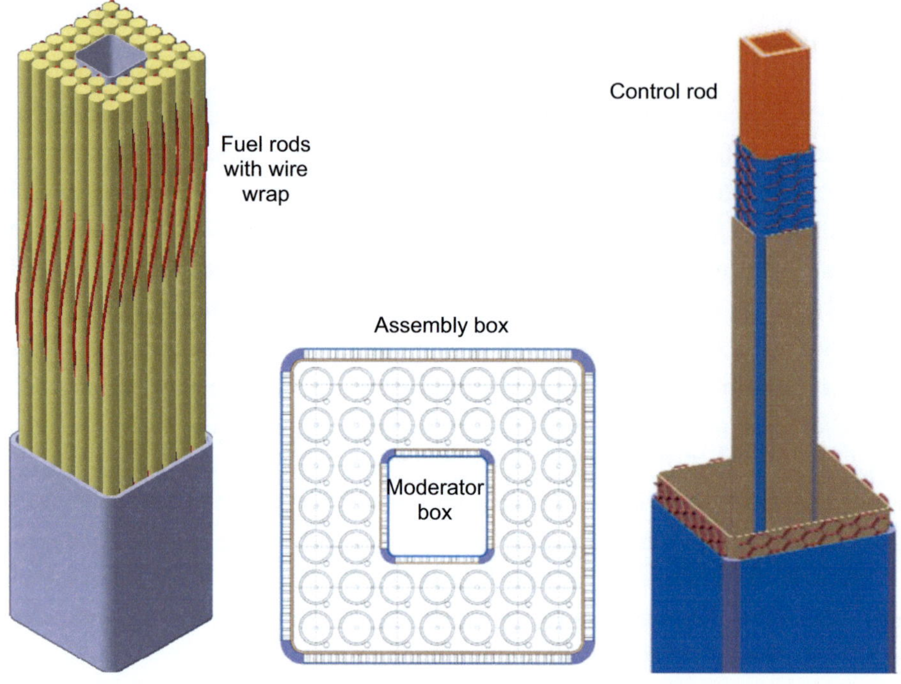

Fig. 2.3 Assembly design with wire wrapped fuel rods (left) and honeycomb structures of the assembly and moderator box (right). A square control rod is inserted from top

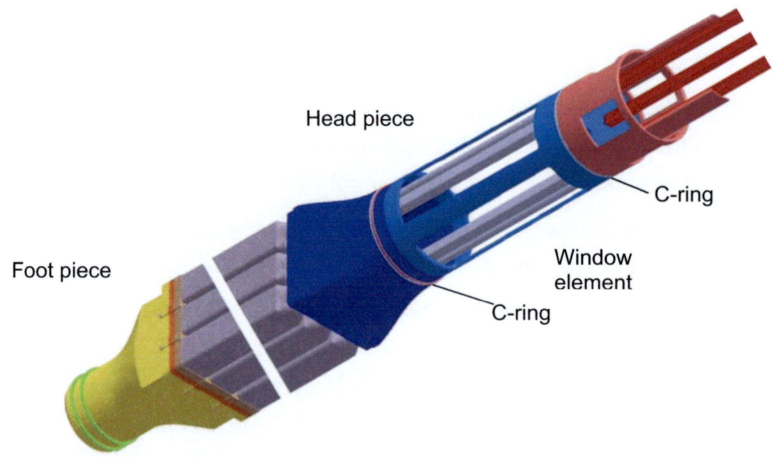

Fig. 2.4: Assembly cluster design with head and foot piece; control rods are running inside 5 of the 9 moderator boxes, inserted from the top [7]

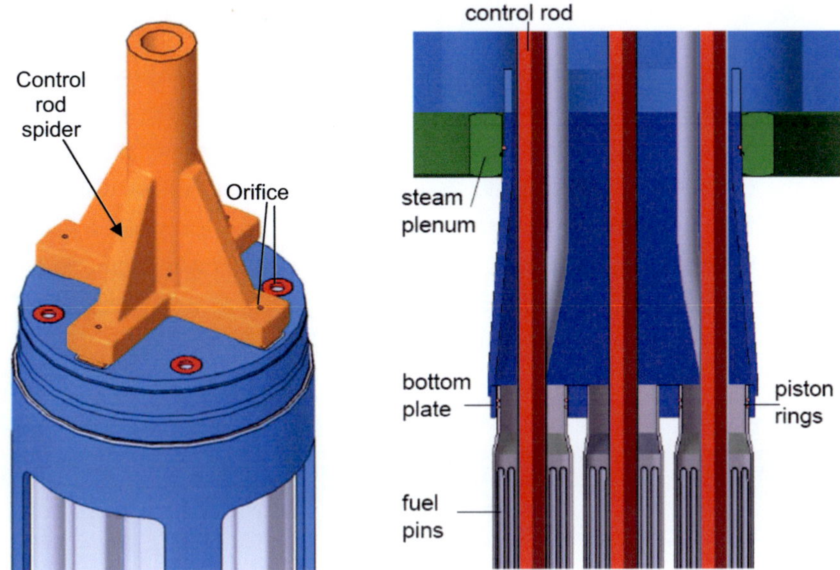

Fig. 2.5: Head piece of an assembly cluster; cut through assemblies and water boxes [7]

Different control rod designs, e.g. cruciform rods and square tubes, filled either with AgInCd or B_4C with different B-10 enrichment, have been studied by Schlagenhaufer et al. [9]. The best shut down reactivity, i.e. the reactivity change if all these control rods are inserted, could be achieved with square tubes, as shown in Fig. 2.3, right, if 90% enriched B-10 is filled as B_4C between two square, stainless steel tubes. The shut-down reactivity of these 5 control rods inserted into a cluster with perfectly neutron reflecting boundaries has been predicted by Schlagenhaufer et al. [9] as given in Tab. 2.2. Fresh fuel with 5% U-235 enrichment and 90% enriched B-10 has been assumed for this MCNP study.

Coolant density [g/cm³]				
0.7770	0.2075	0.2083	0.2237	0.2251
0.05878	0.3005	0.3041	0.3266	0.3289

Tab. 2.2: Shut-down reactivity of different control rods, absorber: 90% B-10 as B_4C [9]

The foot piece is designed with an upper plate (brown), an insert (orange) and a diffuser (yellow), which becomes a nozzle in case of the first superheater, as shown in Fig. 2.6. All but the central assembly box of the cluster are welded with the upper plate. The central assembly box is bolted with 4 screws of the size M10 with the upper plate, instead. Sealing between the upper plate and the central assembly box is provided by sealing lips in both parts which are pressed together by the bolts. All central moderator boxes inside the assemblies are welded with the head piece. Their lower ends are extended with cylindrical tubes which are inserted into the insert of the foot piece, as shown in Fig. 2.6. Piston rings avoid leakage at these joints. The insert includes a channel system which guides the moderator water horizontally to the exit holes in the diffuser, where the moderator water is released to be mixed with the gap water surrounding the assemblies. The outlet orifices can be used to adjust the mass flow rate through the water boxes. Openings for the vertical steam flow, surrounding the insert, are designed as large as possible to minimize pressure losses. The insert of the foot piece and the diffuser are welded together to avoid leakage of cold moderator water into the superheated steam.

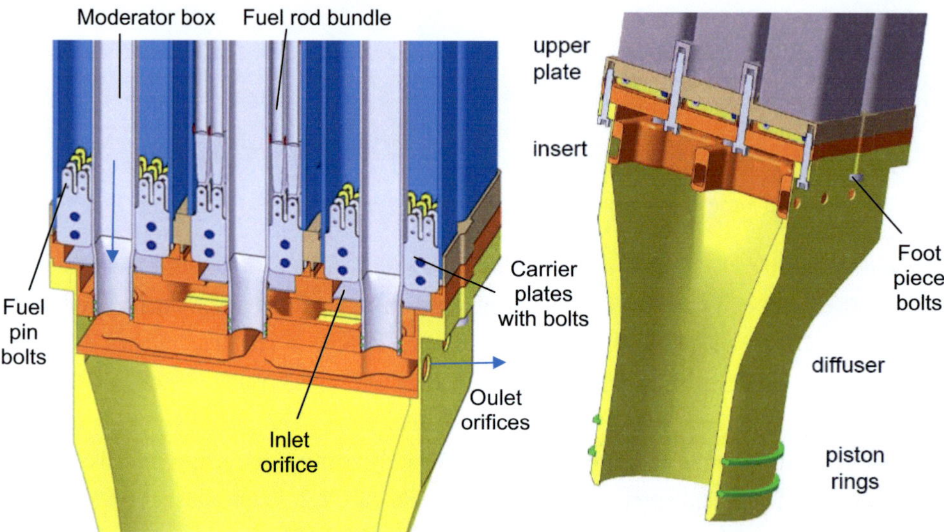

Fig. 2.6: Foot piece design; cut through assemblies (left) and cut between assemblies (right) [7]

After insertion of the fuel bundles into the assembly boxes, the insert and the upper plate of the foot piece are bolted together with 8 screws of size M8. Now the insert keeps the fuel bundles in place. The upper plate and the insert have sealing lips at their contact surfaces which are pressed together by the bolts. The total arrangement is completely separating moderator and steam mass flows without any significant leakage.

If necessary, a fuel pin can be replaced during revisions with the following disassembly steps. The cluster is turned upside down, the 12 bolts of the foot piece are opened, and the damaged fuel bundle is pulled out. After opening the 4 bolts of the carrier plates, each row of fuel pins can further be disassembled by drilling out the fuel pin bolts of the damaged pin. The cluster is reassembled in the reverse order.

Inlet orifices for the coolant mass flow are required only at the evaporator inlet to avoid density wave oscillations, as will be discussed below. They are connected with the insert and can be removed, if necessary, by replacing the entire foot piece. A common orifice for each cluster, indicated in Fig. 2.7 in red, which is needed to adjust the cluster mass flow rate to the cluster power in a certain core position, is remaining on the core support plate when the cluster is taken out to shuffle it to a new position.

Mixing chambers above and underneath the core have been designed by Fischer et al. [7] and updated by Koehly et al. [10] to enable a downward flow of moderator water inside the moderator boxes, but an upward flow in the gaps between the assembly boxes. Details of the core inlet chamber and the lower mixing chamber are shown in Fig. 2.7. Three assembly clusters, namely an evaporator cluster, a first superheater cluster and a second superheater cluster (from right to left) are shown exemplarily.

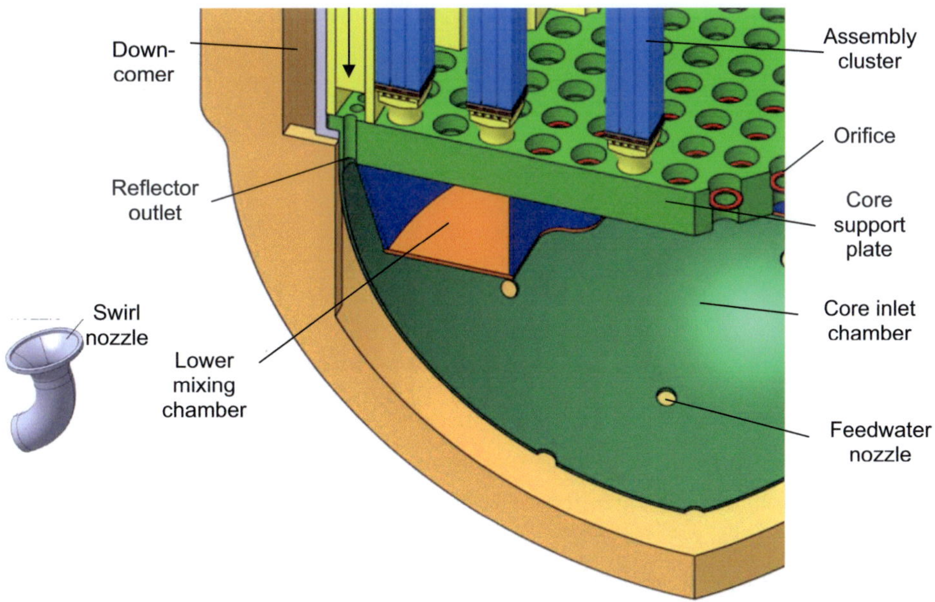

Fig. 2.7: Core inlet chamber and lower mixing chamber underneath the core support plate [10]

Mixing in the core inlet chamber is provided by the jets of feedwater nozzles, which supply 50% of the feedwater running downwards through the downcomer, while the remaining fraction of 50% is cooling the reflector. This latter mass flow is entering the core inlet chamber through penetrations of the core support plate at the reflector outlet.

The lower mixing chamber inside the core inlet chamber is hermetically separated from it by a welded construction of plates. The coolant enters this chamber after release from the first superheater assemblies before entering the second superheater. Coolant mixing in the lower mixing chamber may be enhanced by swirl nozzles, as indicated by Fischer et al. [7]. They are welded with the core support plate at each superheater 1 outlet to cause a vortex ring in this chamber.

The steam plenum is mounted over the window elements of the head pieces as shown in Fig. 2.8. It rests on support struts of the reactor pressure vessel to be aligned with the extractable steam lines. A bellow between steam plenum and reflector minimizes bypass flows. Moderator water, rising up between the assembly boxes, is entering the reflector through openings at the top of the reflector.

Fig. 2.8: Steam plenum mounted over the head pieces of assembly clusters [10]

Inside the steam plenum, welded steel walls are separating an upper mixing chamber from a core outlet chamber, as shown in blue in Fig. 2.9. Moreover, flow obstacles in form of plates with reduced height, shown with gray colour, enhance mixing as shown by Wank [11]. Vertical tubes (blue) are connecting the upper and lower plates of the steam plenum and stiffen the construction. Moreover, they provide penetrations for the core instrumentation arranged between the assembly boxes at the corner of several assembly clusters. Sealing of the steam plenum against ingress of moderator water is avoided (or at least minimized) by C-rings around the head pieces, as shown in Fig. 2.4, and around the steam lines.

The core instrumentation will be described in Chapter 3.5 below.

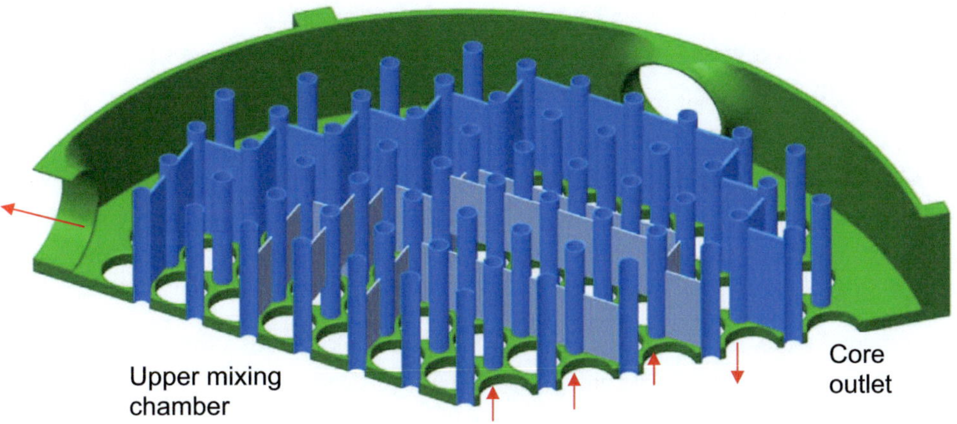

Fig. 2.9: Separating (blue) and mixing walls (gray) inside the steam plenum [11]

2.4 Analysis of the core power distribution

Like with boiling water reactors, the core power distribution is significantly influenced by the coolant density distribution which is responsible for neutron moderation, together with the moderator water inside the water boxes and between the assembly boxes. The coolant density, in turn, is decreasing by the fissile power so that both, the coolant and moderator heat up and the core power distribution must be analysed iteratively to yield a consistent, coupled solution.

Waata et al. [12] applied such an iterative procedure by coupling the neutronic code MCNP with the thermal-hydraulic sub-channel code STAFAS to predict the power and coolant density profile of a single HPLWR assembly defined by Hofmeister et al. [13]. The coolant densities and fuel temperatures of the STAFAS analyses were given as an input to MCNP, which predicted the local power of each fuel rod to be given back by MCNP again. Under-relaxation factors were needed to achieve convergence of this method. As a first result, the corner rods of the assembly required a lower U-235 enrichment to avoid overheating.

The method has been extended by Maraczy et al. [14] to the total core, as described above, by running iteratively the neutronic code KARATE, originally developed by Hegedüs et al. [15] for a hexagonal fuel lattice, with the one-dimensional, parallel channel thermal-hydraulic code SPROD, developed by Yamaji et al. [16] for supercritical water cooled reactors. KARATE, an assembly wise nodal analysis based on the response matrix method for two energy groups, is resolving 39 axial layers for one quarter of the core. SPROD solves the energy and mass equations using heat transfer correlations of Watts and Chou [17] and Jackson and Hall [18] and pressure drop correlations of Rehme [19] for wire wrapped fuel rods.

Maraczy et al. [14] report how a parameterized diffusion cross section library was prepared for the HPLWR assembly with the MULTICELL neutronic transport code. The parameterized cross sections were used by the KARATE program system, which was verified by comparative Monte Carlo calculations.

Five different types of fuel assembly clusters have been selected for a first loading of the core with fresh fuel having a U-235 enrichment of up to 7% and with 2% Gd poisoning of 4 rods next to the mid of each assembly box wall as indicated in Tab. 2.3. A lower enrichment by 1% pt. has been chosen for the corner rods to avoid local overheating there. The control rods of 7 clusters are assumed to be inserted for burn-up compensation as indicated with red color in Fig. 2.10 for ¼ of the core. Note that the corner assemblies of each cluster are not equipped with control rods. Black solid lines separate superheater 1 from evaporator and superheater 2 assemblies. Dashed white lines indicate the size of assembly clusters, and white numbers refer to the cluster types as defined in Tab. 2.3. The radial form factor, i.e. the individual assembly power normalized with the average assembly power of each heat-up step, shown in Fig. 2.11, results in a rather uniform distribution up to around 1.25 with the exception of 2 assemblies in superheater 2 which exceed a form factor of 1.3. The local assembly power is reduced at control rod positions, causing a higher power in assembly clusters without control rods, namely by 5.6% in evaporator clusters and by 4% in clusters of the 1st superheater.

Cluster type	U-235 Enrichment [w/o]			No. of Gd doped pins	Gd_2O_3 content [w/o]
	Basic	Corner	Gd doped		
1	4.0	3.0	-	-	-
2	5.0	4.0	-	-	-
3	6.0	5.0	5.5	4	2.0
4	7.0	6.0	6.5	4	2.0
5	3.0	2.0	-	-	-

Tab. 2.3: Initial enrichment of the fresh fuel

The moderator density is greater than 700 kg/m³ in the water rods inside assemblies and greater than 600 kg/m³ in the gap volume between assembly boxes thanks to the thermal insulation of box walls and thanks to the high mass flow of moderator water, as plotted in Figs. 2.12 and 2.13.

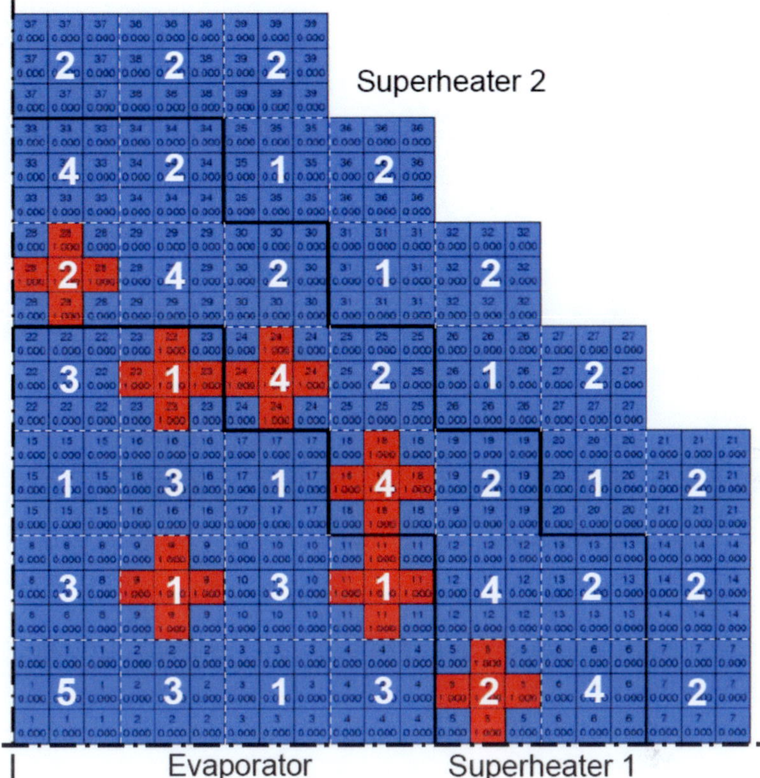

Fig. 2.10: Position of control rods inserted for reactivity compensation of fresh fuel and cluster types according to Tab.2.3 [41]

The radial power distribution becomes worse, however, with increasing burn-up. Starting from fresh fuel as described above, an equilibrium cycle has been predicted by Maraczy et al. [20] with KARATE and SPROD using up to 4 cycles of 1 year each. Different from conventional light water reactors, the new assembly clusters are not inserted at the outer core positions but rather at the outer positions of the evaporator region, whereas older assembly clusters are preferred in the superheater 2 region to achieve the envisaged power distribution. The pattern of clusters of different age is shown in Fig. 2.14. The small upper numbers 4 and 6 refer to the cluster types used to replace the fuel as defined in Tab. 2.4. A 1% lower enrichment has been used in the lower half of the core to account for the coolant density profile. Again, four fuel rods have been doped with Gd. The shuffle scheme of this equilibrium core is shown in Tab. 2.5. The cluster positions in Tab. 2.5 refer to the small upper numbers 1 – 39 in Fig. 2.10.

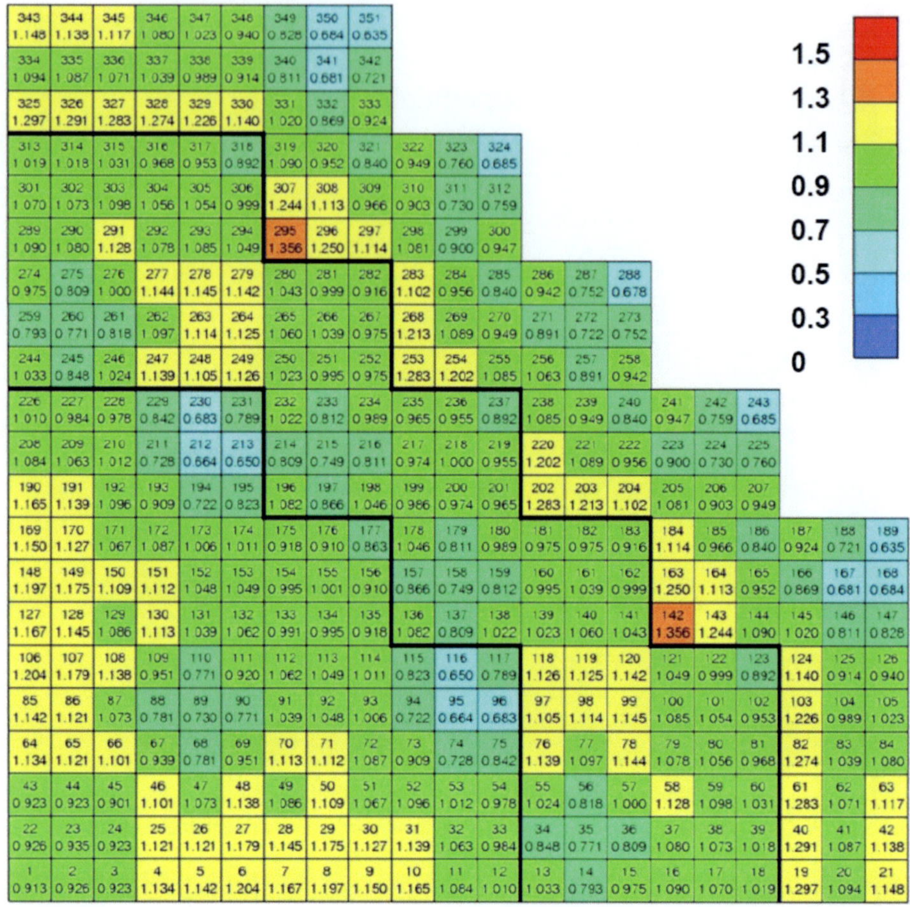

Fig. 2.11: Radial form factor of assemblies with fresh fuel, normalized with the average power of each heat-up section, control rods inserted [14]

Cluster type	Axial segment [cm]	^{235}U enrichment [w/o]			No. of Gd doped pins per assembly	Gd$_2$O$_3$ content [w/o]
		Basic	Corner	Gd doped		
4	000.00-204.61	6.0	5.0	5.5	4	2.0
	204.61-420.00	7.0	6.0	6.5	4	2.0
6	000.00-204.61	6.5	5.5	6.0	4	3.0
	204.61-420.00	7.0	6.0	6.5	4	3.0

Tab. 2.4: Enrichment of assembly clusters to replace fuel in the equilibrium cycle

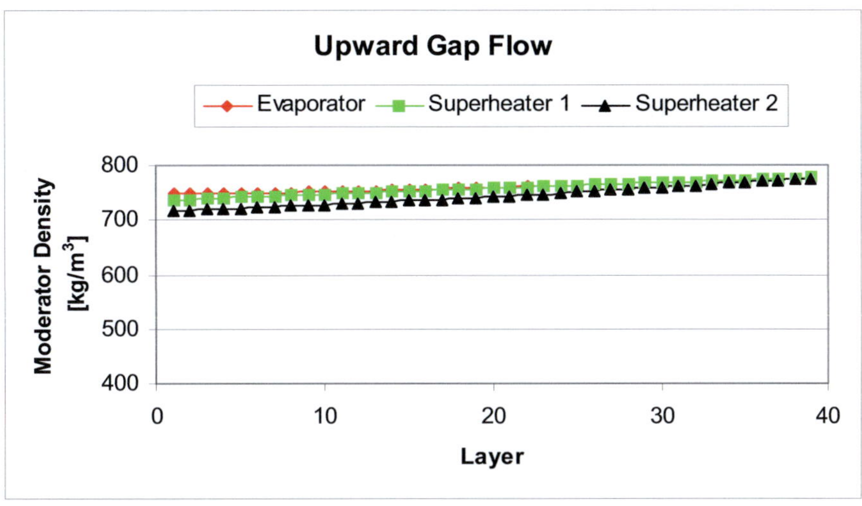

Fig. 2.12: Average moderator density in the water rods of each core segment [14]

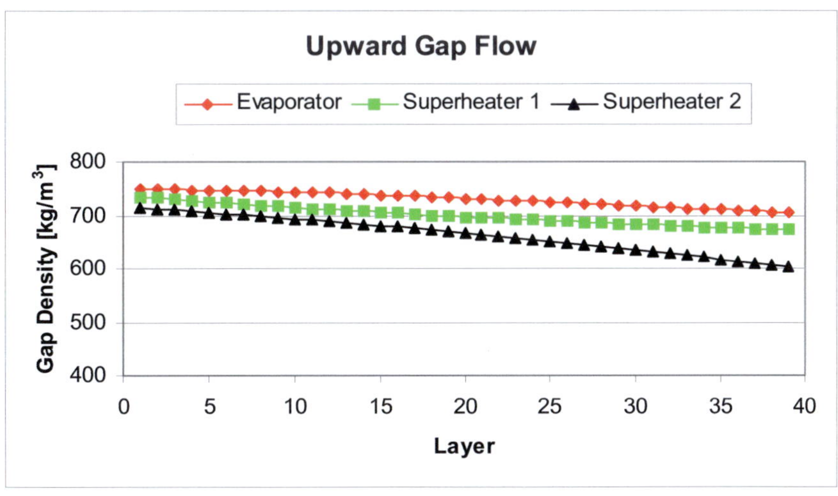

Fig. 2.13: Average moderator density in the gap between assembly boxes for each core segment [14]

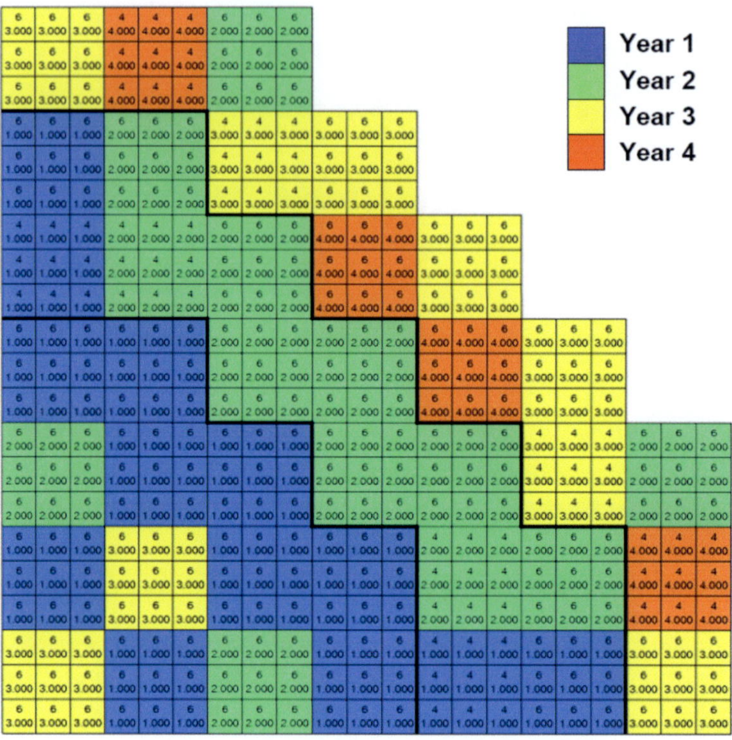

Fig. 2.14: Age of clusters in the equilibrium core [41]

Residence time [cycles]	Cluster type	Position in the nth cycle			
		1st	2nd	3rd	4th
2	6	10	18	-	-
	6	16	24	-	-
	6	6	3	-	-
	6	33	15	-	-
3	6	17	25	1	-
	6	4	19	32	-
	6	22	30	9	-
	6	11	13	7	-
	6	23	34	37	-
4	6	2	21	27	26
	6	8	39	36	31
4	4	5	12	20	14
	4	28	29	35	38

Tab. 2.5: Shuffle scheme of the equilibrium cycle

The radial power distribution (i.e. the total assembly power scaled by the average assembly power) at the beginning of the burn-up cycle and at the end of the burn-up cycle are shown in Figs. 2.15 and 2.16, respectively. As envisaged, the power is highest in the evaporator, where more than 1400 MW of the total thermal power of 2300 MW is supplied at the beginning of the cycle, whereas the 2nd superheater is producing only around 100 MW. During burn-up, the evaporator power is decreasing as fuel is faster consumed there, so that only 1300 MW are produced there at the end of the cycle, whereas the power of the 2nd superheater is increasing to around 150 MW.

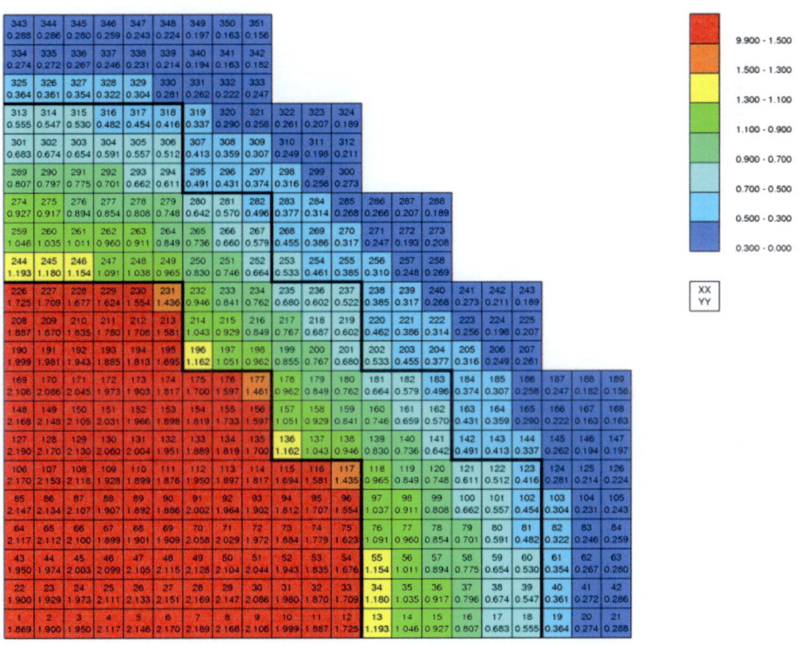

Fig. 2.15: Relative core power distribution at the beginning of an equilibrium cycle (BOC) [51]

The burn-up distribution at the beginning and the end of the equilibrium cycle at the mid-plane of the core is shown in Figs. 2.17 and 2.18, respectively. We find a fast fuel consumption in the evaporator, but a slow one in the 2nd superheater.

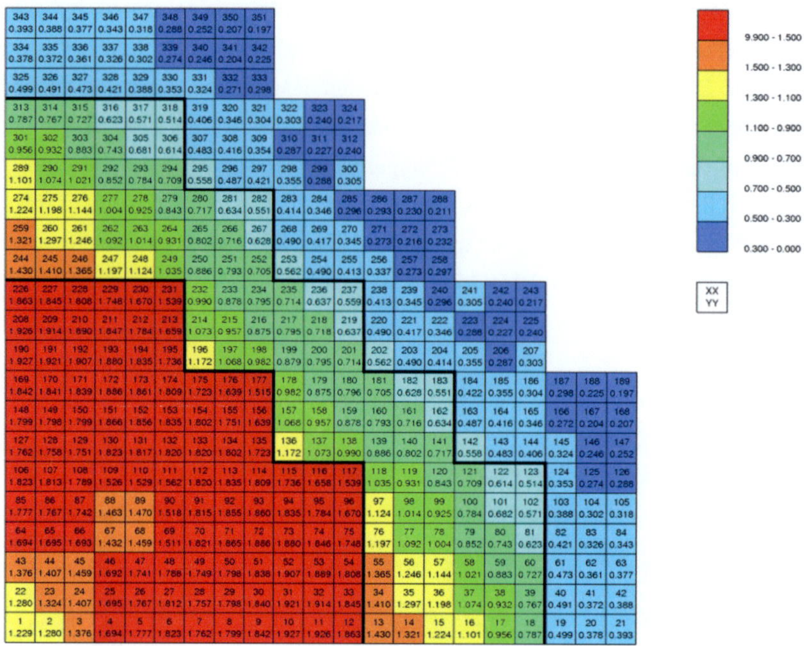

Fig. 2.16: Relative core power distribution at the end of an equilibrium cycle (EOC) [52]

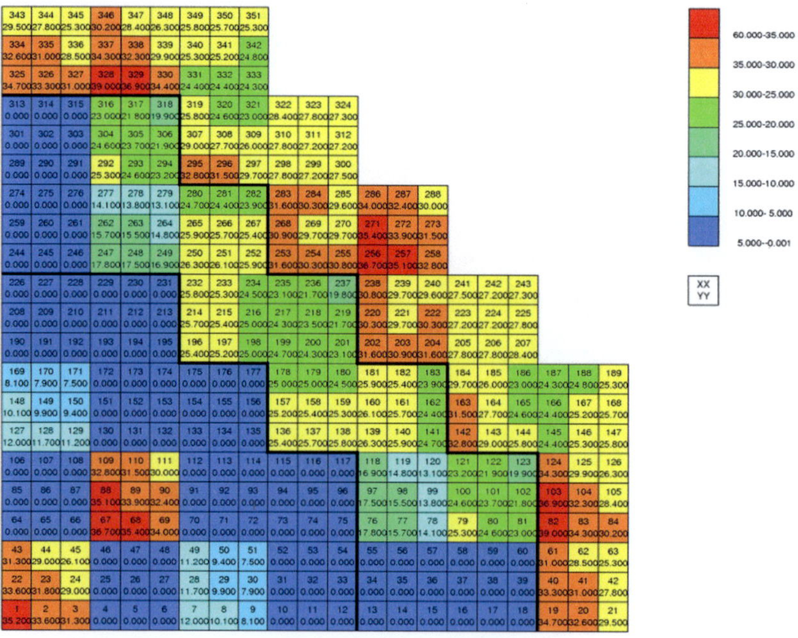

Fig. 2.17: Radial burn-up distribution of assemblies at the beginning of the equilibrium cycle in MWd/kg$_{HM}$ at the core mid-plane [52]

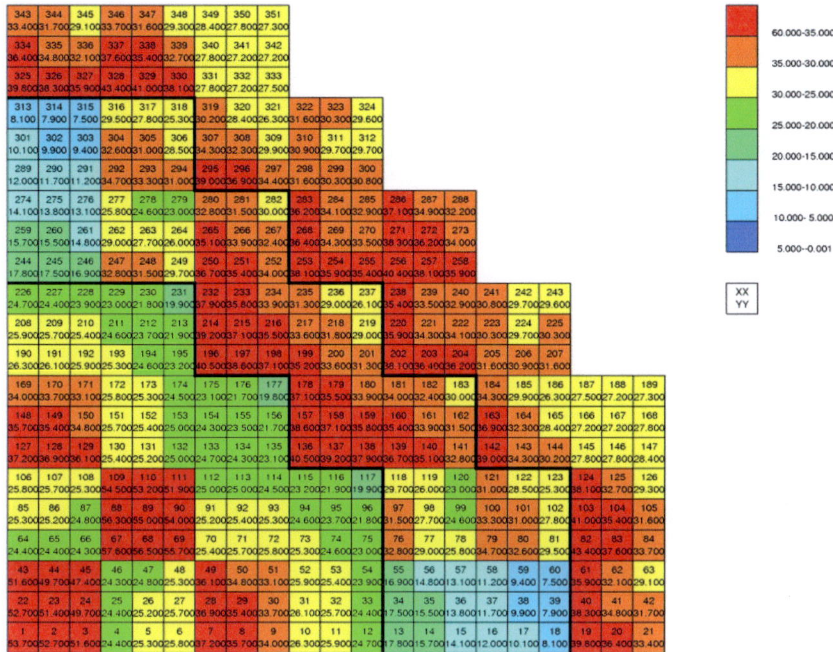

Fig. 2.18: Radial burn-up distribution of assemblies at the end of the equilibrium cycle in MWd/kg$_{HM}$ at the core mid-plane [51]

The neutron multiplication factor k$_{eff}$ with fully withdrawn control rods is reduced to 1 within about 1 year at full power, as shown in Fig. 2.19, which will require to open the reactor again. Besides the effect of burnable absorbers, control rods can be applied to compensate the excess reactivity during the cycle as no boric acid reactivity control is allowed.

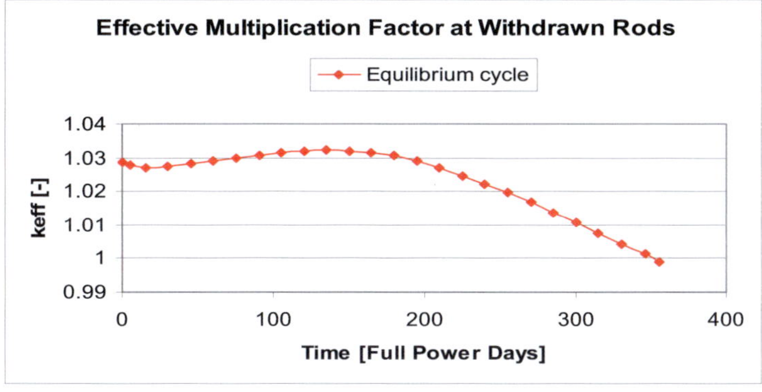

Fig. 2.19: Effective neutron multiplication factor during an equilibrium burn-up cycle, control rods withdrawn [52]

51

We read from Fig. 2.18 (lower numbers in each assembly) that the peak burn-up reached more than 50 MWd/kg$_{HM}$, but the average discharge burn-up is significantly less. Tab. 2.6 yields a total average of 32.5 MWd/kg$_{HM}$ only.

Residence time [cycles]	Cluster type	Number of clusters	Discharge burn-up [MWd/kg$_{HM}$]
2	6	16	30.89
3	6	20	35.80
4	6	8	30.62
4	4	8	29.56
3 (average)	6-4	52	32.53 (average)

Tab. 2.6: Discharge burn-up of different cluster types of the equilibrium core [52]

Higher burn-up could be achieved with a higher initial enrichment. Assuming an enrichment of assembly clusters as given in Tab. 2.7, we get a cycle length of 330 days and an average residence time of the assemblies of 6.5 years, such that 12 clusters will be discharged after 6 years and 12 clusters after 7 years. The average discharge burn-up will be over 60 MWd/kg$_{HM}$.

Axial segment [cm]	^{235}U enrichment [w/o]			No. of Gd doped pins	Gd$_2$O$_3$ content [w/o]
	Basic	Corner	Gd doped		
000.00-204.61	9.0	8.0	8.5	4	3.0
204.61-420.00	9.5	8.5	9.0	4	3.0

Tab. 2.7: Enrichment of assembly clusters to exceed an average discharge burn-up of 60 MWd/kg$_{HM}$

If we divide the individual assembly power, Figs. 2.15 and 2.16, by the average power of each heat up step, we get the radial form factors of the evaporator and both superheaters. Figs. 2.20 and 2.21 show these radial form factors at the beginning and at the end of the equilibrium cycle, respectively. Control rods are assumed to be fully withdrawn. According to the design strategy of Schulenberg et al. [5], these radial form factors should be less than 1.25. Instead, however, we get form factors of more than 1.5 in the first superheater and more than 1.8 at the beginning of cycle in the second superheater.

To compensate local power peaks, at least partially, the hotter assemblies could be cooled with a higher coolant mass flow, which can be achieved by optimization of the inlet orifice of the clusters shown in Fig. 2.7. The pressure drop coefficients K of each cluster and the total pressure drop along the coolant flow path from the evaporator inlet to the 2nd superheater outlet are shown in Tab. 2.8. The resulting coolant mass flow rates of all assemblies are shown in Figs. 2.22 and 2.23 at beginning and end of cycle, respectively.

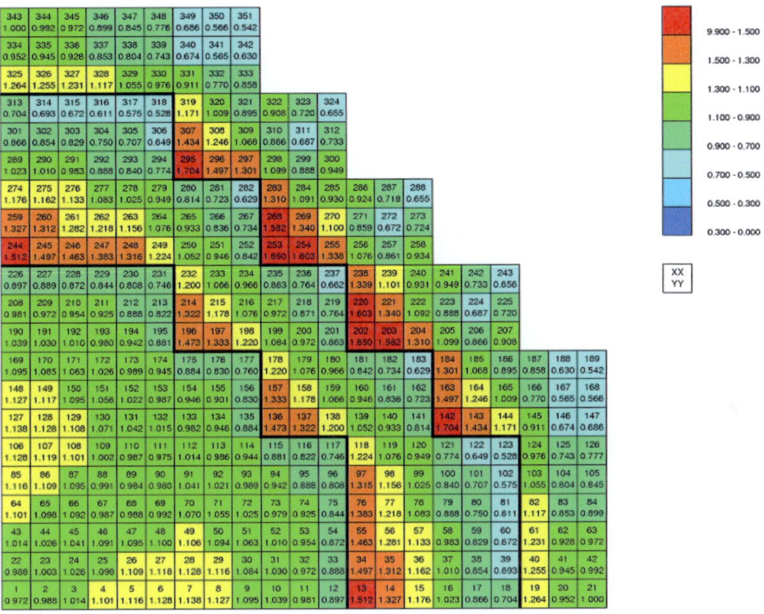

Fig. 2.20: Radial form factors at the beginning of the equilibrium cycle (BOC) [52]

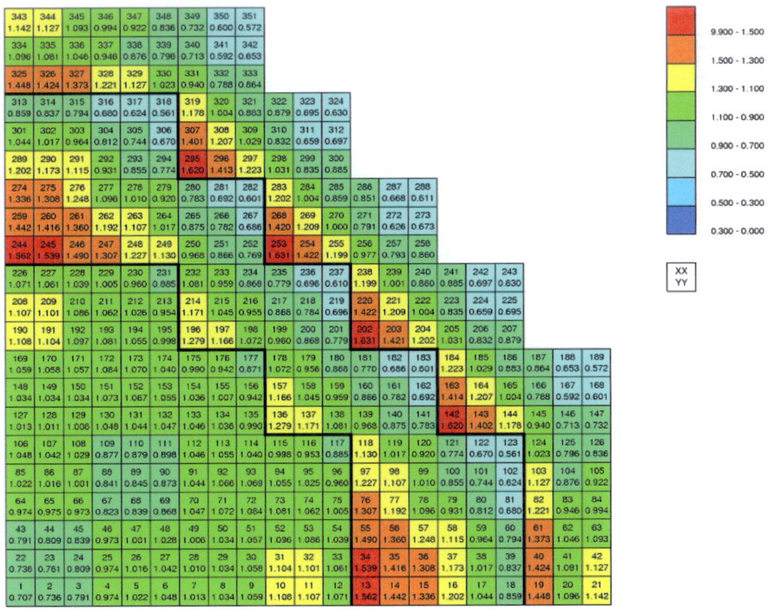

Fig. 2.21: Radial form factors at the end of the equilibrium cycle (EOC) [41]

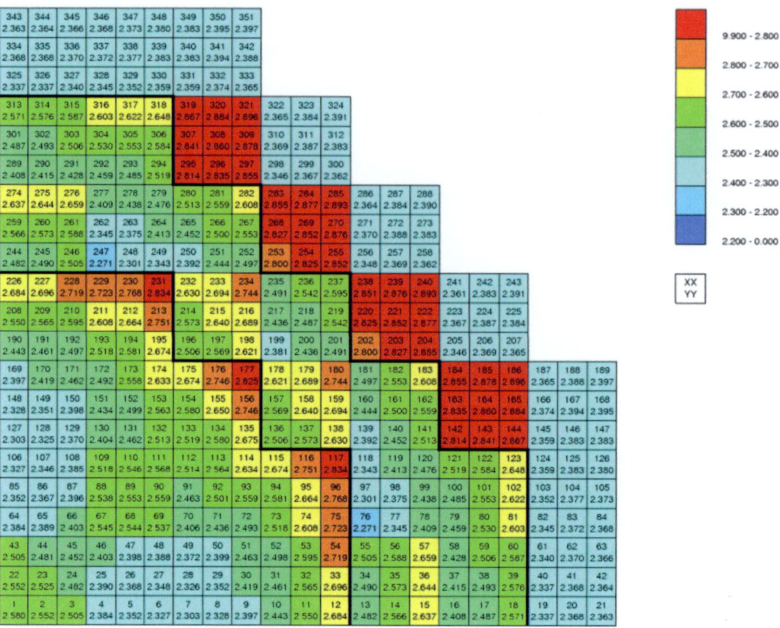

Fig. 2.22: Coolant mass flow rate of assemblies in kg/s at the beginning of cycle [52]

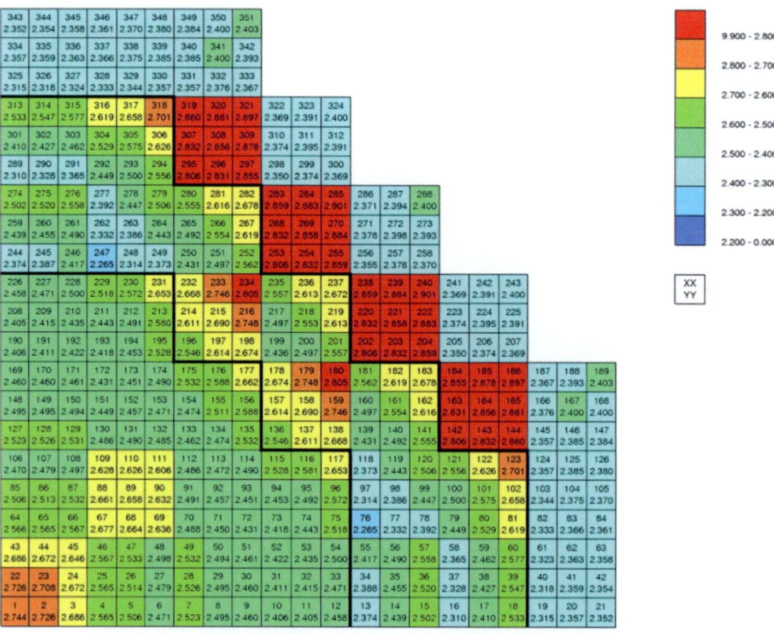

Fig. 2.23: Coolant mass flow rate of assemblies in kg/s at the end of cycle [51]

	Hot cluster number	K_{bulk}	K_{hot}	Δp_{BOC} [bar]	Δp_{EOC} [bar]
Evaporator	-	11	-	1.21	1.13
Superheater 1	5,18,24,28	6	0	2.18	2.00
Superheater 2	20,26, 31, 35	9	0	2.78	2.72
Total				6.17	5.85

Tab. 2.8: Pressure drop coefficient of cluster orifices and pressure drop along the coolant flow path. Cluster numbers refer to Fig. 2.10

The axial power profile of each heat up step, averaged radially over each heat up step, is shown in Fig. 2.24. The reduced enrichment in the lower core region is avoiding an excessive linear power.

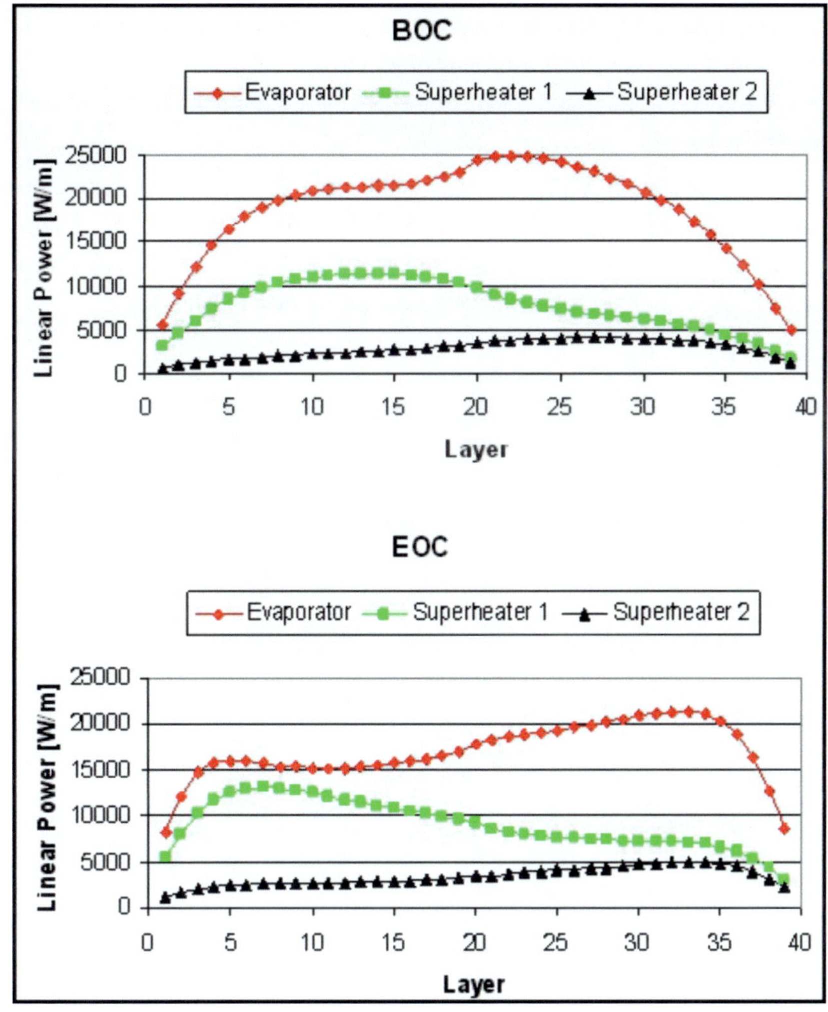

Fig. 2.24: Axial linear power profile, radially averaged, of each heat up step [52]

2.5 Local peaking factors

The steeper gradients of the neutron flux in the second superheater are not only causing larger power differences of the assemblies of a cluster, but also of individual fuel rods inside these assemblies. Monti [21] succeeded to estimate the power of each fuel rod of the core by analyzing first the global flux distribution with the neutron transport code ERANOS [22] coupled with the thermal-hydraulic code TRACE by Murray and Staudenmeier [23], which he multiplied then with the power distribution of a single assembly analyzed with MCNP5 for a given neutron flux. An exemplary analysis of such a pin power reconstruction technique has been performed for a core with fresh fuel of uniform enrichment of 5%, except the corner rods having 4% only, but its results can also be taken to estimate the local peaking factor of the core described above.

The hottest coolant temperatures can be found in the 2nd superheater, next to the first superheater, e.g. in assemblies 295, 253, 202 or 142 in Figs. 2.20 and 2.21. There, the pin power reconstruction should yield fuel rod power factors similar to the ones shown in Fig. 2.25 where the hottest fuel rod is having 18% more power than the average.

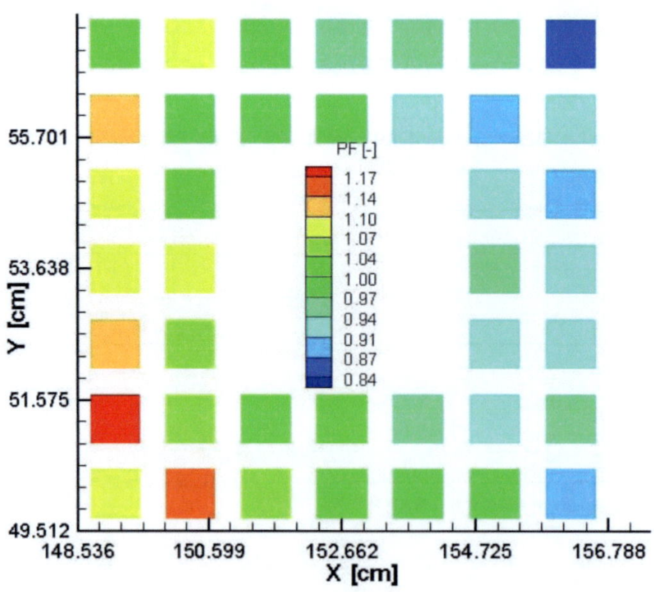

Fig. 2.25: Power peaking factors (PF) of individual fuel rods of the hottest superheater 2 assembly [21]

These power distributions were taken then as input to a sub-channel analysis with MATRA [24], assuming perfectly insulated box walls for simplicity and wire wraps as spacers. Fig. 2.26 shows the temperature distribution in the sub-channels of the upper part of such assemblies, where the peak temperatures are only about 10°C hotter at the exit than the average temperature, thanks to the excellent mixing of the wire wraps. This corresponds with an enthalpy peaking factor of 1.08 to be multiplied with the average enthalpy rise of the second superheater. Similar results have been obtained by Himmel et al. [25], who assumed a radial power gradient of 20% inside a superheater 2 assembly and predicted a maximum difference of the coolant temperature to its average at the assembly outlet of 25°C. This corresponds with an enthalpy peaking factor of 1.12. Unfortunately, this local peaking factor has the same cause as the radial power profile described above, so that both factors need to be multiplied.

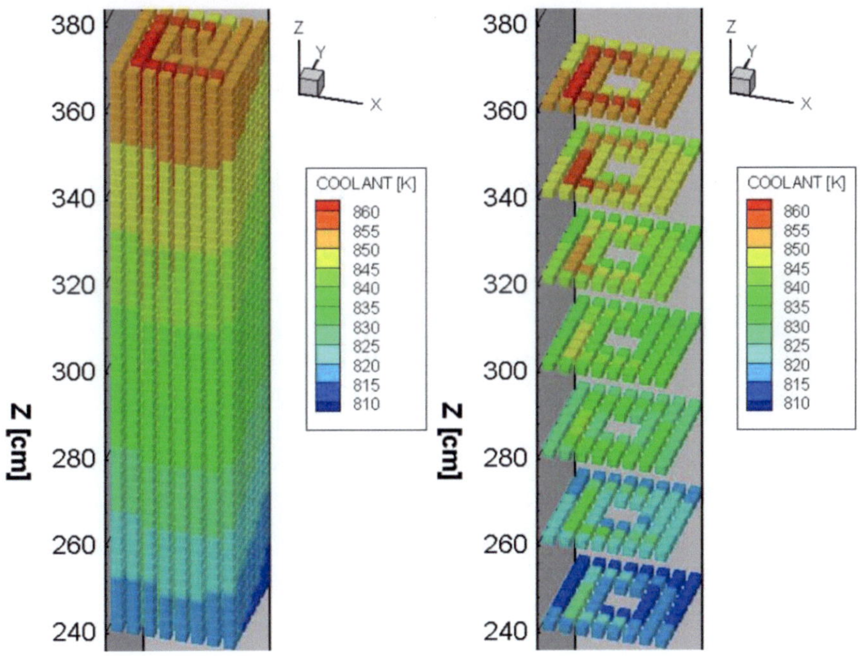

Fig. 2.26: Sub-channel coolant temperatures of the hottest assembly of superheater 2 after pin power reconstruction of the local power profile [21]

The hottest fuel temperature is expected to occur in the evaporator, where the coolant temperature is still rather cold, but the assembly power is highest. The hot spot of the fuel centerline temperature and the maximum linear heat rate were predicted by Monti [21] to occur in assembly number 1, assuming fresh fuel with uniform enrichment as described above. The peak power and temperature profiles of each fuel rod are shown in Fig. 2.27.

A similar pin power reconstruction of evaporator assembly number 6, Figs. 2.20 and 2.21, by Maraczy et al. [20] resulted in the maximum linear power of the core of 365 W/cm at the beginning of an equilibrium cycle, which is reducing to 305 W/cm at the end, resulting in fuel centerline temperatures of 2390K and 2156K respectively. These numbers are slightly smaller because fresh fuel has been omitted in the central fuel assembly clusters.

Other causes for local peaking factors are clusters with inserted control rods and non-uniformities due to Gd burnout at the end of a burn-up cycle. As an example, we show in Fig. 2.28 an evaporator cluster with inserted control rods at the beginning of a burn-up cycle. All fuel rods contained fresh, 6% enriched fuel except the corner rods which had 5% enrichment, and 2.5% Gd was added to the corner rods of each assembly for burn-up compensation. The 2D burn-up analysis was performed by a coupled Monte Carlo transport and burn-up calculation for half an assembly using symmetry boundary conditions, as explained by Nabbi and Bernnat [26]. 20 actinides and 85 fission products were regarded explicitly. The densities and temperatures of moderator and coolant, the fuel temperature and the assembly power were kept constant during burn-up.

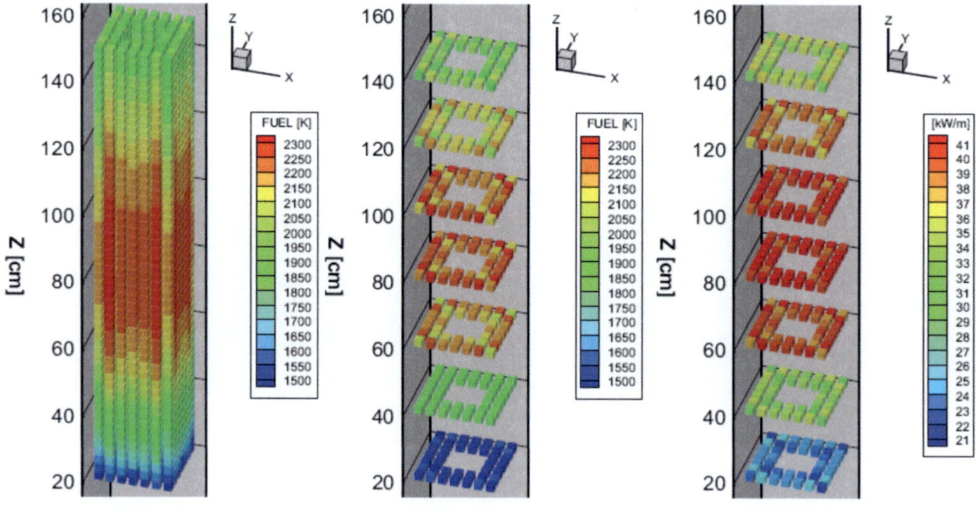

Fig. 2.27: Fuel centerline temperatures and linear heat rate of fuel rods of the hottest evaporator assembly [21]

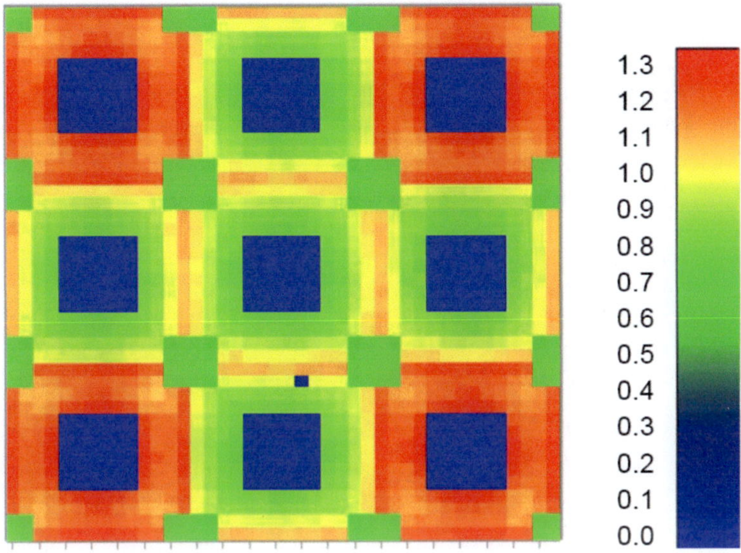

Fig. 2.28: Local power peaking factors of an evaporator cluster with inserted control rods at the beginning of a burn-up cycle; corner rods contain 2.5% Gd_2O_3 [51], [41]

As clusters with control rods are usually not the hottest ones, confirmed by Fig. 2.11, the maximum peaking factor of 1.3, indicated in Fig. 2.28, does not need to be multiplied with the radial form factor, but evaporator clusters without control rods receive an average overpower of 5.6% instead.

Fig. 2.29 shows the same cluster with extracted control rods after 20GWd/t_{HM} burn-up. Now, the initially Gd-doped corner rods reach a power peaking factor of almost 1.3. Thanks to coolant mixing, the enthalpy peaking factor is expected to be lower, as will be described next.

These exemplary results show that a local enthalpy peaking factor of 1.15, as assumed in the initial thermal-hydraulic analysis of Schulenberg et al. [5], could be reached by careful optimization, even though this number does not seem to be conservative regarding these results.

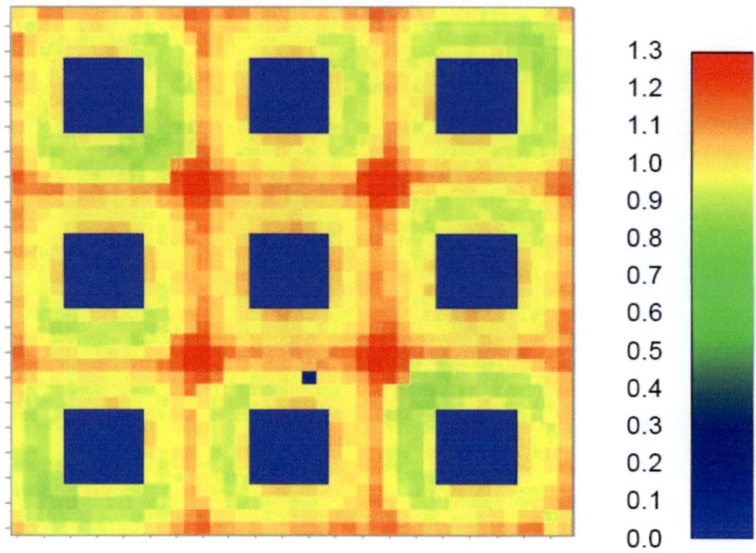

Fig. 2.29: Local power peaking factors of an evaporator cluster with Gd burnout after a burn-up of $20\,GWd/t_{HM}$; control rods extracted [52], [41]

2.6 Coolant mixing

As a measure to manage the high enthalpy rise of the coolant in the core, an effective coolant mixing inside assemblies and between each heat up step is a key requirement of this core concept. Mixing between sub-channels inside assemblies has been studied with CFD analyses by Himmel et al. [27]. A single wire wrapped around each fuel rod, which had already been applied successfully to sodium cooled fast breeder reactors in the past, turned out to be an effective mixing device which works well in both flow directions. Therefore, it allows using the same assembly design in the evaporator as well as in both superheater sections. The CFD results of single sub-channels were taken to model the flow in the entire assembly with a coarse grid. Different from conventional sub-channel analyses, Himmel et al. [28] were using the commercial CFD code STAR-CD for full rod bundle analysis by applying minor modifications to it. The reaction forces obtained by the local CFD analyses were simply added as body forces to the momentum equation to cause the same global flow structure as with the detailed analysis. Instead of writing a dedicated code with numerical solver routines and post-processing tools for sub-channel analyses, they applied the optimized Graphical User Interface already provided in STAR-CD. Thus, a smooth transition to full three-dimensional modeling of the fluid flow inside rod bundles was enabled with the same code system, just by refining the spatial discretization.

As an example, Fig. 2.30 shows the coolant temperature distribution of an assembly in the second superheater, with and without wire wraps, exposed to a horizontal power gradient of 20%. The effectiveness of the wire is most pronounced in the assembly corners which tend to overheat without the wire: instead, the maximum temperature difference in the outlet cross section is decreased from 60°C to 16°C by the wire.

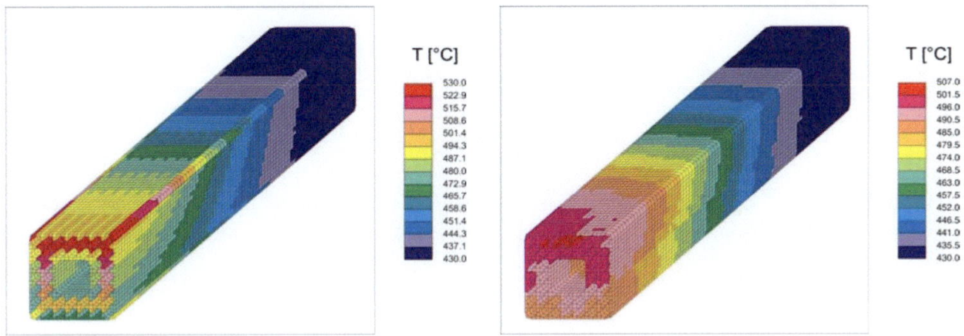

Fig. 2.30: Coolant temperatures in an assembly of the second superheater without (left) and with (right) mixing induced by the spiral wires [28]

The flow structure with wire wraps becomes more obvious in a CFD analysis of the entire assembly flow, modeled for a single axial wire pitch of 200 mm by Kiss et al. [29]. Fig. 2.31 shows the streamlines of an unheated assembly section modeled with CFX-11. The cross flow velocities are strongly affected by the sweeping effect of the wires. A clockwise and counterclockwise flow near the assembly box and near the moderator box can be indentified which purges the corners quite effectively. The normalized velocity ratio of a heated case was almost the same as the unheated case which means that heating did not affect the cross velocity profile. Enhanced mixing is shown to be caused by the directed flow along the wires and by the enhanced turbulence caused by the wires. Most sub-channels exchange mass not only with their next neighbors, but also with sub-channels that are further away.

The effectiveness of coolant mixing with wire wraps has also been confirmed by sub-channel analyses with C3CLM. The abbreviation C3CLM stands for COBRA IIIC Liquid Metal. It is described by Pütz [30] and refers the COBRA IIIC as originally developed by Rowe [31]. Fig. 2.32 shows the effect of a single fuel rod of a superheater 2 assembly having 20% more power than the others (indicated in red). The coolant temperature at the outlet of this assembly is plotted along the path indicated in the figure. Obviously, the temperature is only increased in the direct vicinity of the hot rod. The maximum change of the local temperature is just 1°C.

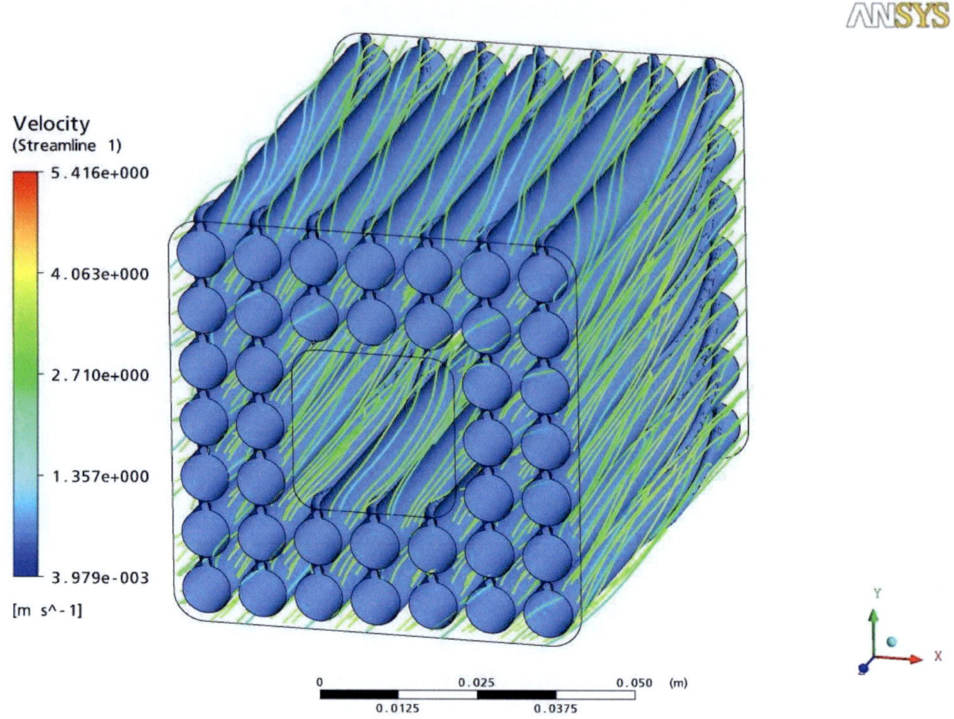

Fig. 2.31: Flow structure in an assembly with wire wraps enhancing mixing [29]

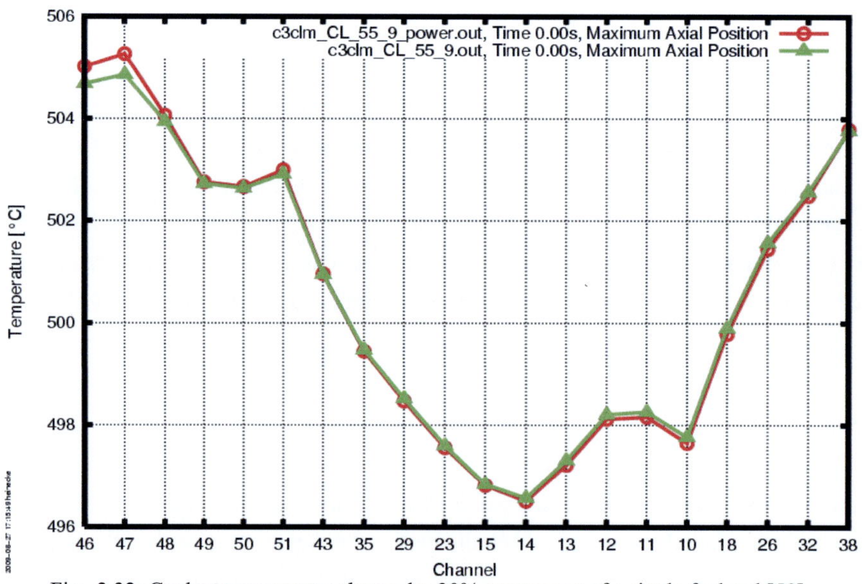

Fig.. 2.32: Coolant temperature change by 20% overpower of a single fuel rod [50]

Coolant mixing in the upper mixing chamber, Fig. 2.1, has been studied by Wank et al. [32]. The initial design concept of the upper mixing chamber was guiding the coolant from the evaporator outlets directly to the inlets of the first superheater without any significant mixing. Therefore, Wank [33] tried several alternative options to enhance coolant mixing in a numerical design study with STAR-CD. The analysis was simplified as an isothermal flow, and mixing of hot streaks from any evaporator cluster outlet was modeled by individual tracers which were added to each outlet. Thus, the mixing quality could simply be checked by evaluating the tracer concentrations from each outlet at the inlet of each superheater cluster. His best result was achieved with additional walls that were welded into the mixing chamber as shown in Fig. 2.9. The gap between these walls and the steam plenum was causing jets which improved mixing significantly.

Fig. 2.33 shows a cut through the upper mixing chamber with concentrations of tracers added to different cluster outlets of the evaporator. We see that the peak concentrations at the superheater inlets are less than 39% of the inlet concentrations, even if a superheater inlet is situated just next to an evaporator outlet. As a drawback, the pressure drop of the upper mixing chamber increased to 50 kPa.

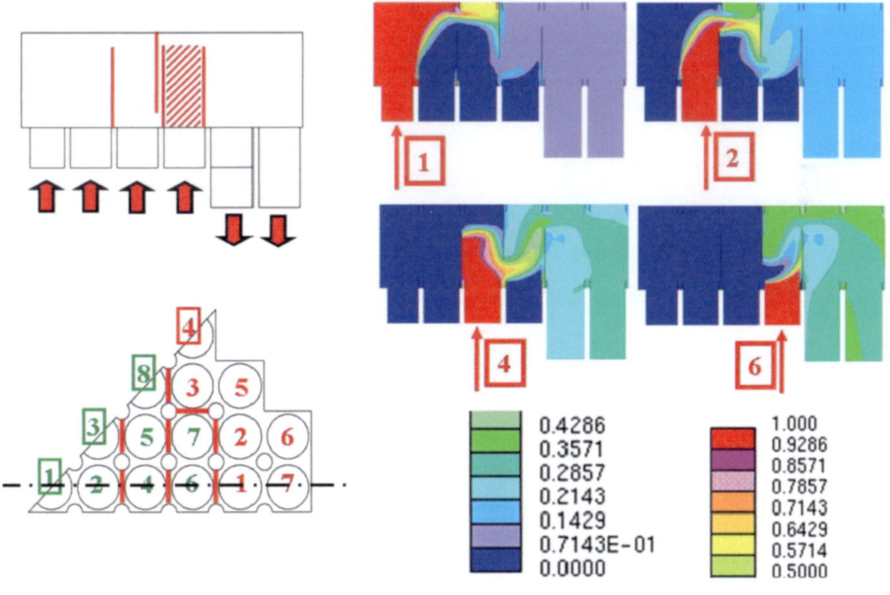

Fig. 2.33: Mixing of tracers released from outlets of evaporator clusters in the upper mixing chamber as predicted by Wank et al. [33]. Relative units

If we zoom further into the details of coolant mixing, which is released from single assemblies of a cluster in a cross flow of other clusters, Möbius et al. [34] predicted even that less than 16% of the inlet cluster concentration is arriving as peak concentration at the neighboring cluster inlet. Smaller jets are thus mixing better than the entire cluster flow.

The optimized design of the upper mixing chamber, Fig. 2.9, was analyzed then by Wank [33] with respect to mixing of a realistic outlet coolant temperature distribution predicted by Maraczy et al. [14] for a core with fresh fuel optimized for a first core loading. Some of the control rods were assumed to be inserted in the evaporator for reactivity compensation in this analysis, causing colder jets of the coolant released from the evaporator in clusters 3 and 7 (Fig. 2.33). The predicted coolant temperature distribution at the cluster outlets and inlets is shown in Fig. 2.34. The coolant temperature varies only by 1.5°C at the superheater inlets which, however, is also due to the temperature range close to the pseudo-critical point. The coolant enthalpy varies by 47 kJ/kg (max. – min.) at the superheater inlets, which is less than 15% of the inlet enthalpy spread.

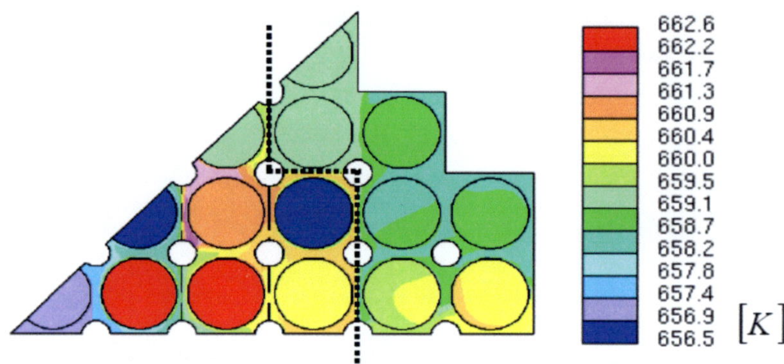

Fig.2.34: Coolant temperature distribution at cluster outlets and inlets of the upper mixing chamber [33].

A similar study has been performed for the lower mixing chamber, Fig. 2.7. Different from the upper mixing chamber, which is filled with numerous head piece structures, moderator boxes and tubes, the lower mixing chamber is empty and jets can be used as mixing devices. For this purpose, the coolant released from the foot pieces of the first superheater is passing outlet nozzles which are directed such that the flow in the lower mixing chamber is forming an annular vortex. The swirl nozzle is indicated in Fig. 2.7. Wank [33] reports how the nozzles shall be oriented. Like with the upper mixing chamber, concentrations of tracers released from the first superheater outlet are shown at the second superheater inlets in Fig. 2.35.

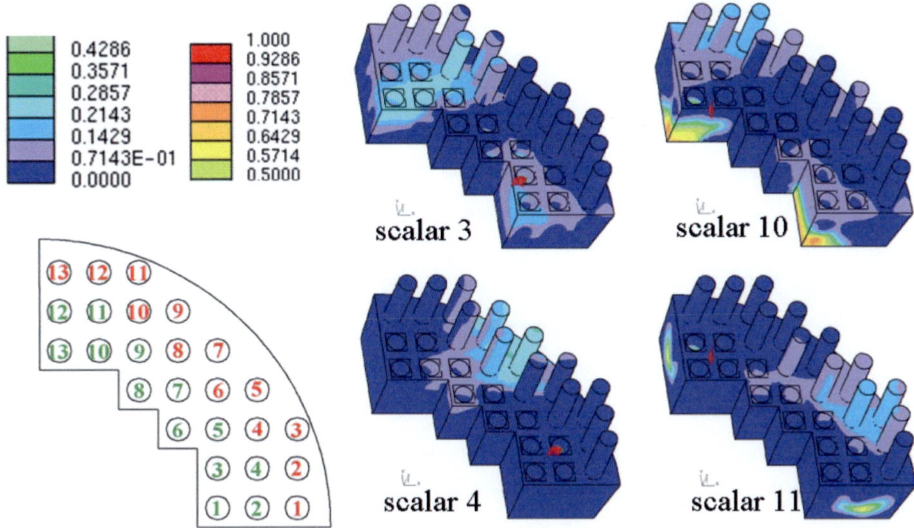

Fig. 2.35: Mixing of tracers released from outlets of superheater 1 clusters in the lower mixing chamber, Wank [33]. Relative units

Applying again the core power distribution of Maraczy et al. [14], we get a coolant temperature distribution as shown in Fig. 2.36. The maximum temperature spread at the second superheater inlet is 7K, corresponding with an enthalpy spread of 40 kJ/kg (max. – min.). This is 26% of the enthalpy differences at the first superheater outlets.

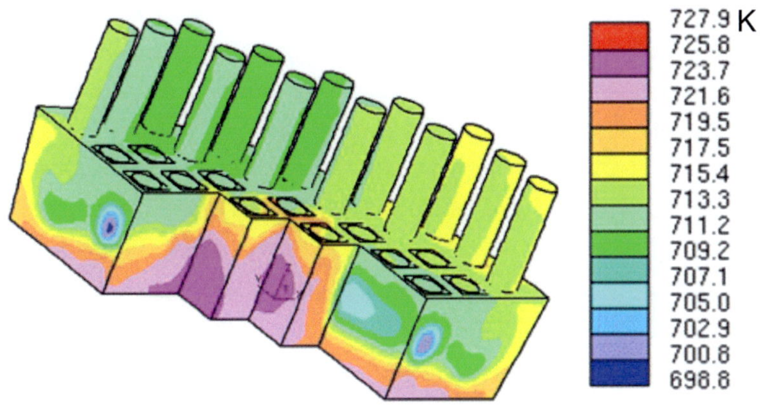

Fig. 2.36: Coolant temperature distribution in K in the lower mixing chamber [33]

2.7 Moderator flow

Before entering the fuel assemblies, 50% of the coolant is used first as moderator water in the moderator boxes inside the assemblies and in the gaps between the assembly boxes. In a first approach, both moderator flows were assumed to be supplied from the top of the core, from where they should run downwards to be mixed with the downcomer flow in the core inlet chamber underneath the core. However, Kunik et al. [35] predicted a flow reversal in the large gap volume between the assembly boxes causing a rather non-uniform moderator density distribution. The gap volume was simplified as a porous medium and local CFD analyses were performed to determine the hydraulic resistance coefficients in their analyses.

As a flow reversal with larger density variations of moderator water would influence the core power distribution significantly, the flow path was changed to a supply of all moderator water through the head pieces into the moderator boxes only, Figs. 2.3 to 2.5. Afterwards, this water was released through the foot pieces, Fig. 2.6, to rise upwards through the gaps between the assemblies. This stable flow configuration was studied by Kunik et al. [36] applying the same porous media approach again. Fig. 2.37 shows the temperature distribution of moderator water between the assembly boxes. The moderator water supplied through the foot pieces is hotter in outer core positions, because it had initially been pre-heated to higher temperatures in the moderator boxes of the superheaters. Now the temperature distribution is stable, and the maximum coolant temperature is still well below the pseudo-critical temperature, keeping its high density.

At the top of the core, the moderator water enters the radial reflector as indicated in Fig. 2.8. It has been designed with large, vertical water channels as shown in Fig. 2.38 (left) to optimize the neutron reflection, flattening the power profile in the second superheater. The flow structure inside 2 of these channels has been checked by Kunik et al. [36] with CFD analyses using STAR-CD. The temperature distribution, Fig. 2.38 (right), shows a small heat up in the reflector of less than 3°C, which increases steadily from top to bottom without any flow reversal.

The analyses have been repeated at 60% load and 60% mass flow rate, assuming the same coolant temperature inside assemblies as with full load. Results are confirming that the flow structure is not altered at lower velocities.

Flow reversal in the moderator boxes can be avoided by use of inlet orifices of these flow channels in the head piece, as shown in Fig. 2.5. The orifice diameter is chosen such that the pressure drop is larger than the expected hydrostatic pressure difference in the moderator boxes, which are due to different heat up in evaporator and superheater assemblies. A thermal insulation of these boxes is helping to keep these temperature differences low. The orifices have been designed here for a minimum mass flow rate of 50% of the nominal case, which will cause an important constraint for the start-up procedure. The unavoidable gap between a control rod and a moderator box must be minimized accordingly.

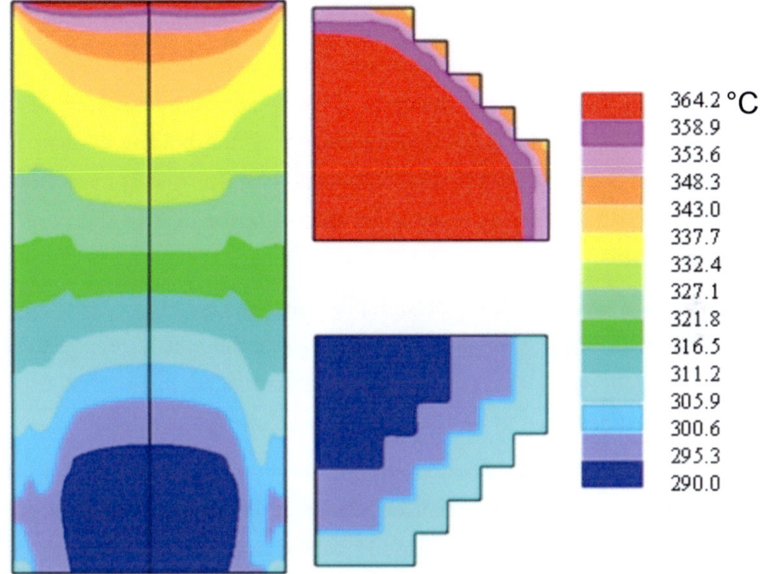

Fig. 2.37: Temperature distribution of moderator water in °C in the gaps between assembly boxes, view from the core symmetry planes (left), from top and from below (right), Kunik et al. [36]

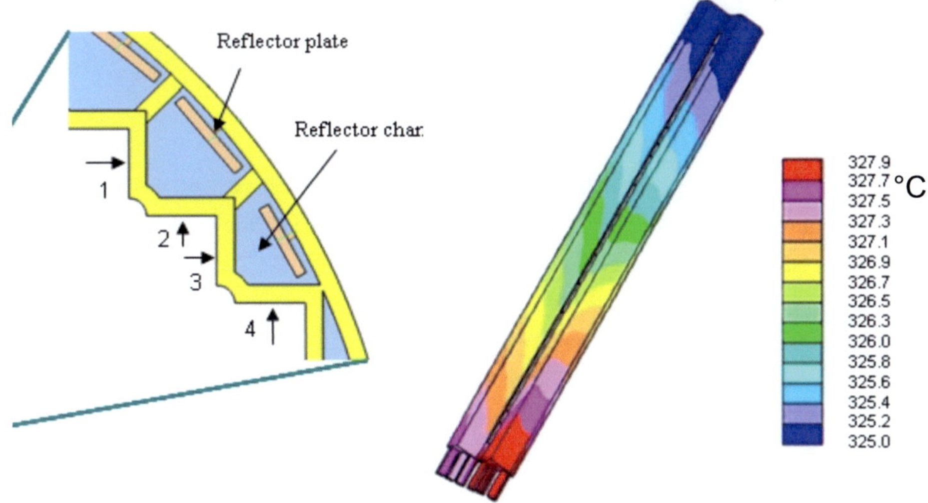

Fig. 2.38: Reflector design with water channels (left) and water temperatures predicted inside, Kunik et al. [36]

2.8 Stability issues

2.8.1 Density wave oscillations

Flow stability problems are not only limited to the moderator flow. A stability problem which is well known from boiling water reactors (BWR) is the occurrence of density wave oscillations. It is caused by the large density change of the boiling coolant in the core, in particular if the local coolant pressure drop increases with decreasing mass flow. The coolant density ratio in the HPLWR changes by more than a factor of 8 in the core, i.e. it is even higher than in a BWR.

Stability analyses of the coolant flow through the three pass core have been studied by Ortega Gomez [37]. Like with BWR, Ortega Gomez shows that the most effective measure to avoid density wave oscillations in the core is the installation of orifices at the inlet of fuel assemblies. These orifices need to be customized for a hot fuel assembly.

In case of a BWR, the operation point of the average heated fuel assembly should correspond to a decay ratio less than 0.5 for a single channel density wave oscillation, and a decay ratio less than 0.25 should correspond to the coupled thermal-hydraulic/ neutronic density wave oscillation (DWO). Furthermore, the whole operation range, also including hot fuel assemblies, should be in the linear stable region of the stability map. Note that the decay ratio DR is defined as

$$DR = exp\left[-2\pi \frac{Re(\Lambda)}{|Im(\Lambda)|} \right] \qquad (2.1)$$

where $Im(\Lambda)$ and $Re(\Lambda)$ are the imaginary part and the real part of the leading mode at given operation parameters. Hence, the decay ratio should be below 1 for the whole operation range. The stability guidelines of BWRs were extended for the fuel assemblies of HPLWR heat-up components (evaporator, superheater I and superheater II). A uniform heat-up has been assumed exemplarily. The resulting decay ratios of the linear stability analysis for fuel assemblies of all three HPLWR heat-up components are listed in Table 2.9, assuming that neither inlet nor outlet orifices are foreseen. The cases shown here are the single channel DWO, the in-phase DWO, and the out-of-phase DWO.

It can be seen that the average and even the hot fuel assemblies of the superheaters fulfill the stability criterion for all three types of DWO without applying any orifice. In contrast, however, average fuel assemblies of the evaporator have a decay ratio larger than 0.25 at normal operation parameters for the in-phase and out-of-phase DWO. Furthermore, hot fuel assemblies of the evaporator would operate in the linear unstable region. Thus, while the fuel assemblies of the superheaters do not need additional inlet flow restriction, all fuel assemblies of the evaporator stage must be equipped with inlet orifices.

	Single channel DWO	In-phase DWO	Out-of phase DWO
Average evaporator	0.034	0.5	0.52
Hot evaporator	2.25	1.19	1.18
Average superheater I	0.018	0.035	0.039
Hot superheater I	0.041	0.061	0.065
Average superheater II	0.010	0.022	0.041
Hot superheater II	0.013	0.043	0.048

Tab. 2.9: Decay ratios for HPLWR three pass core heat-up components without orifices [37].

Figs. 2.6 and 2.7 show that the assembly clusters have two sets of orifices. Each fuel assembly cluster has an orifice at the inlet of the diffuser of the foot piece, Fig. 2.7. This orifice adjusts the mass flow so that the coolant of clusters at different positions in the evaporator reaches a uniform coolant exit temperature. However, Ortega Gomez [37] shows that these cluster orifices do not cause a stabilizing effect. The coolant density may be oscillating out-of-phase in the individual assemblies of a cluster downstream the orifice. Therefore, a second set of orifices needs to be installed at the inlet of each fuel assembly of the evaporator as shown in Fig. 2.6. As a consequence, however, these orifices should be taken out when the cluster is moved from an evaporator to a superheater position.

The stability analysis has been performed for different inlet loss coefficients k_{in}, while an unavoidable exit loss coefficient k_{out} of 2.5 has been assumed. The orifice coefficient of this single orifice has been increased stepwise until the whole operation range of the evaporator is in the linear stable region, and further, until the normal operation point has a decay ratio less than 0.5 for the thermal-hydraulic DWO and less than 0.25 for the reactivity DWOs. Figure 2.39 shows the operation range of the evaporator (green line) in the stability map.

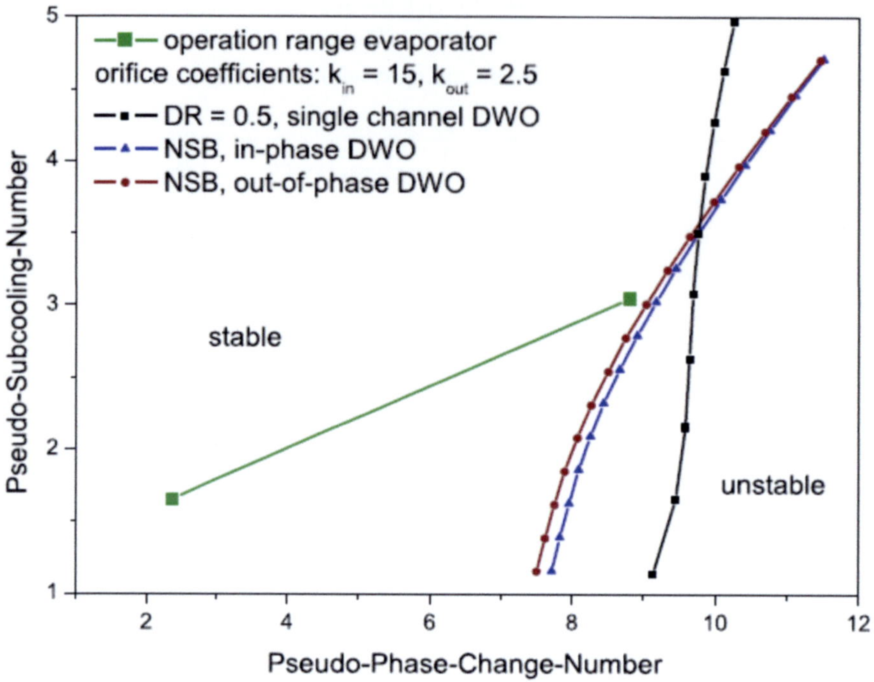

Fig. 2.39: Neutral stability boundary (NSB) is shown for the in-phase (red) and the out-of-phase DWO (blue), while an inlet loss coefficient of k_{in} = 15 and an outlet loss coefficient of k_{out} = 2.5 is applied. The curve of decay ratio 0.5 is given for the single channel DWO. The whole operation range of the evaporator (green line) is in the linear stable region, Ortega Gomez [37]

In this diagram, the pseudo-subcooling-number is defined as:

$$N_{P-SUP} = \frac{(v_{out} - v_{in})z_{PC}}{v_{in}L}$$

(2.2)

where v_{in} and v_{out} are the specific volumes of the coolant at inlet or outlet of the assembly, respectively, z_{PC} is the axial position of the pseudo critical point in the flow channel, and L is the total length of the heated section. The pseudo-phase-change number is defined as

$$N_{P-PHC} = \frac{v_{out} - v_{in}}{v_{in}}$$

(2.3)

With a loss coefficient of 15 for the single inlet orifice, the curves of neutral stability boundary (NSB) for the in-phase (red) and out-of-phase DWO (blue) are at higher Pseudo-

Phase-Change-Numbers than a hot fuel assembly of the evaporator. The same fact is valid for the curve of decay ratio 0.5 for the single channel DWO (black).

The pressure loss due to an orifice at the inlet of the evaporator fuel assemblies is given by

$$\Delta p = K \frac{\rho_{in} u_{in}^2}{2} \tag{2.4}$$

where ρ_{in} is the density of the coolant at $T_{in} = 310\ °C$ (25 MPa) u_{in} is the inlet velocity of the coolant and K is a geometry dependent pressure loss coefficient. For a square-edged orifice, the orifice loss coefficient is given by

$$K = \frac{4}{5}\left[\left(\frac{D_1}{D_2}\right)^4 - 1\right] \tag{2.5}$$

where D_1 and D_2 are the diameter of the flow channel and the reduced diameter by the orifice, respectively. The HPLWR fuel assembly has a cross flow area of 1826 mm^2. Thus, a square-edged inlet orifice should have a reduction of the cross-section area to 408 mm^2.

2.8.2 Flow reversal

While the first superheater is stable with respect to density wave oscillations, even without orifices, we have to expect flow reversal in some superheater assemblies at low mass flow rates due to an unstable stratification of the downward flow. The stability of the coolant flow has been checked systematically for the entire load range with APROS as described in chapter 4.3, comparing the flow direction of a hot channel with the one of a nominal channel. The mass flow control of the reactor is usually such that flow reversal is excluded. Low mass flow rates are unavoidable, however, during some sequences when the reactor is opened and the core is disassembled or during accident scenarios.

The following CFD study of Tiret et al. [38] is addressing an exemplary case of a flow reversal caused by a low mass flow rate to discuss the consequences to be expected in general. A 1/8 sector of the first superheater has been modelled with common plena above and below, simulating the corresponding sectors of the upper and lower mixing chambers. The reactor is assumed to be shut down and the assemblies are heated with a residual heat of 0.66% of the nominal power, i.e. a typical situation during an outage. The coolant is assumed to be single phase liquid with an inlet temperature of 104°C. A cosine axial power profile has been assumed, and the power of the first superheater was uniformly distributed among the assembly clusters.

The computational domain extends from the outlets of evaporator clusters to the inlets of the clusters of the second superheater, as sketched in Fig. 2.40. The inlet velocity is assumed to be 2.5 cm/s, thus a very low velocity which should result in mixed convection phenomena in the first superheater. Flow inside assemblies has been approximated with a porous media approach.

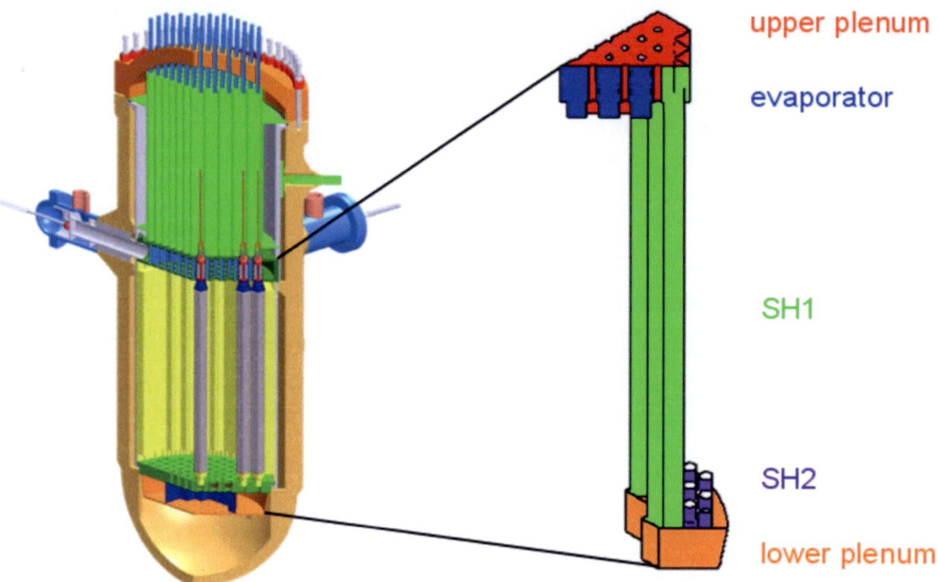

Fig. 2.40: Sector of the first superheater analysed with respect to flow reversal phenomena [38]

Initially, the coolant is assumed to run uniformly downwards through all assembly clusters of the second superheater. Thus, it is heated up from 104°C at the inlet to 129°C at the outlet. After some minutes, the coolant velocity in one of the assembly clusters is slowing down and changing its direction to an upwards flow. The peak coolant temperature of the 1st superheater is plotted in Fig. 2.41 as a function of time, showing a maximum of 160°C after around 540s. Afterwards, the velocity in a second assembly cluster is slowing down and is changing its flow direction as well. We observe a second temperature peak of 168°C at the time 780s. The flow stabilizes at a maximum peak temperature of 165°C. A cut through 3 assembly clusters at the time step at 544s, 568s, 784s and 796s is shown in Fig. 2.42 for illustration. The maximum coolant heat up is thus 2.5 times the average heat up.

The general conclusion of this study is that flow reversal will not be a concern for the core as long as we have enough margin from the cladding temperature limits. Boiling, which was excluded here, will enhance heat removal from the superheater and will accelerate the flow. A non-uniform power distribution will cause the flow reversal to start earlier. As a consequence, flow reversal in the 1st superheater must definitely be excluded under all load conditions, but may be acceptable for residual heat removal with lower temperatures.

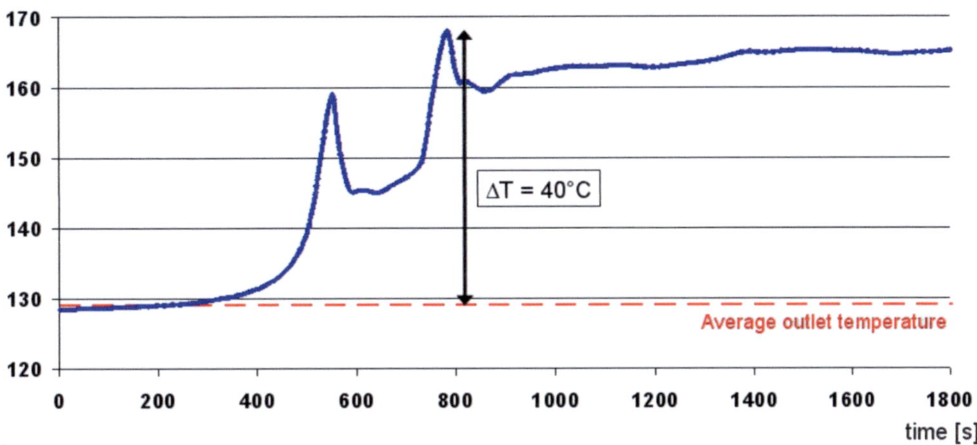

Fig. 2.41: Peak coolant temperatures due to flow reversal in the 1st superheater [38]

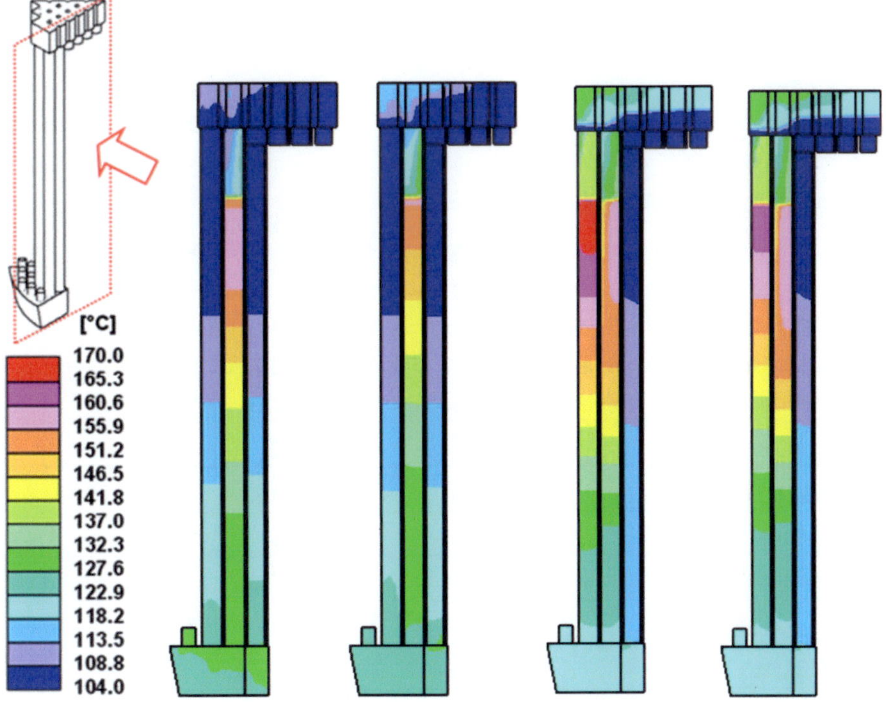

Fig. 2.42: Coolant temperature distribution during the time steps 554s, 556s, 784s and 796s, from left to right [38]

2.8.3 Xenon oscillations

Another stability issue, which was studied in this project, are xenon oscillation of the core power, like those known from conventional light water reactors (LWR). Reiss et al. [39] studied these oscillations with single assembly clusters under HPLWR conditions. The diameter of the core of the HPLWR is around 3.5 meters, while the active height is 4.2 meters. These dimensions are in the range of LWRs where xenon oscillations must be taken care of. On the other hand, due to the large density drop of water after crossing the pseudocritical point, the migration length of the neutrons – which is an important parameter for the stability of the reactor against xenon oscillations – is larger than in current LWRs. The migration length in the HPLWR 3-pass core with a uniform fuel enrichment of 5% and a coarse, but representative temperature distribution was calculated to be 8.79 cm in the z direction (this is about twice as much as in a VVER-1000, where it is about 4.45 cm. This finding and the strong temperature feedback may stabilize the reactor against xenon oscillations.

The characteristic time of the oscillations is several hours, thus a coupled quasi-stationary neutronics / thermal-hydraulics code completed with the xenon poisoning differential equations can predict the extent of these processes. The program system described by Reiss et al [39] is made up of a neutronics part (using the program MCNP) and of a thermal-hydraulics part. Due to the geometrical parameters of the core and the size of the migration length, the calculations are mainly focusing on axial oscillations.

If the thermal-hydraulic feedback is neglected and the coolant and moderator temperatures and densities are given as an input instead, we get the characteristic numbers of the xenon oscillations as listed in Tab. 2.10. Case 1 represents a typical coolant density in the evaporator inpet. Here, the xenon amplitudes are highest. Case 2 represents a 1st superheater assembly, whereas case 3 is typical for a 2nd superheater at its outlet.

	Case1	Case2	Case3
Temperature [K]	553.0 K	656.0 K	773.0 K
Density [kg/m^3]	777.23	390.51	89.79
Migration length (z direction) [cm]	7.027	9.35	12.42
Oscillation period [h]	13.235	10.815	∞
Power peak [MW]	1.07	0.67	0.56
Amplitude	2.6e-6	1.5e-6	-

Tab. 2.10: Effect of coolant temperature on xenon oscillations [39]

A more realistic test case including coolant reactivity feedback is shown in Fig. 2.43 (showing the xenon concentration y in one axial level of an assembly). The oscillations of the xenon concentration y can be approximated as

$$y = y_0 + A \cdot \sin\left(\frac{2\pi}{T}(t - t_c)\right) \cdot e^{bt} \tag{2.6}$$

where t is the time and T the oscillation period. In this test case, the oscillation period is about 15 hours, the stability index b is positive, thus the oscillation is unstable. It should be noted that the oscillations evolve much slower compared with the previous case due to the strong feedback from the coolant density.

The preliminary results indicate that the HPLWR will be unstable against xenon oscillations. Nevertheless, its normal operation can be ensured with proper control equipment (e.g. partly inserted control rods) which is already well-established and will be similar to the ones used in today's large reactors. In the HPLWR at the beginning of the burn-up cycle, some of the control rods are inserted to compensate excess reactivity which make them suitable – besides power control - for xenon oscillation control. At the end of the cycle, some of the control rods will be still inserted because of power control and safety considerations, so they could prevent large oscillations as well. On the other hand, part length control rods could be useful not only for controlling xenon oscillations, but also to fine tune the power distribution during normal operation [53].

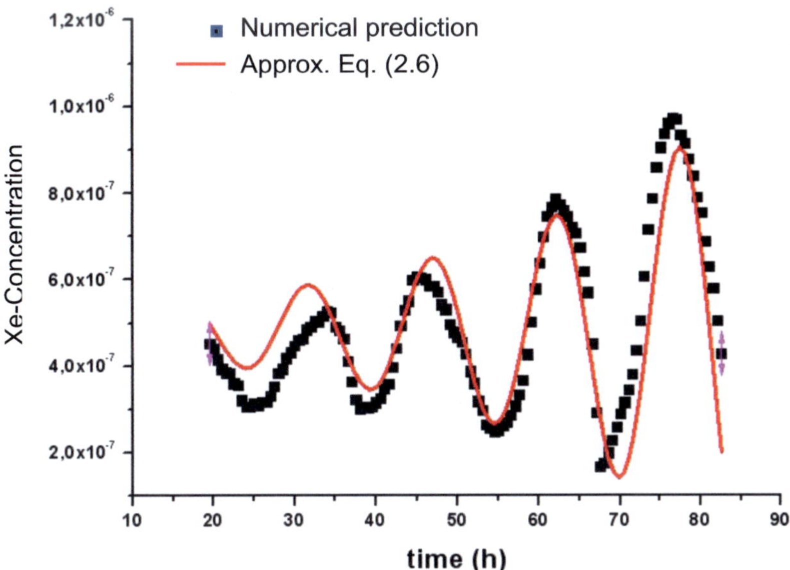

Fig. 2.43: Example of xenon oscillations predicted for a single HPLWR assembly cluster [39]

In order to fully understand the behaviour of the HPLWR 3-pass core against xenon oscillations, a full-core model applying fast computational methods will be required.

2.9   Stresses and deformations of the assembly box

The pressure differences of the coolant from reactor inlet to outlet will be significantly higher than in a BWR. Moreover, temperature differences between coolant and moderator are significantly higher than in a BWR. Both effects are a challenge for the design of the moderator and assembly boxes. Therefore, the assembly box design, Fig. 2.3, has been analyzed in detail with respect to its thermal conductivity, its stresses and its deformations. The smaller moderator box will generally have a higher resistance against the pressure differences and will be guided between the fuel rods, so that it is expected to be less critical with respect to deformations and stresses.

The thermal conductivity of the assembly box has been estimated by Herbell and Himmel [8]. They assumed a stainless steel (SS347) honeycomb structure with 8.6 mm cell size made from 0.2mm sheet material, which is filled with Zirconia of 35% porosity and soaked with water, between 2 stainless steel liners of 0.6 mm thickness on the hotter side and 0.4 mm thickness on the colder side. The average thermal conductivity of the honeycomb layer is 2.1 W/mK so that the total heat transfer coefficient of the sandwich construction is in the order of 1000 W/m^{2}K. The corner pieces are made from solid stainless steel, indicated in blue in Fig. 2.3, center, as finite element analyses of an alternative sandwich design for corner pieces turned out to be too weak to stand the pressure differences.

A short axial cut out of the assembly box was exposed to a pressure difference of 500 kPa. It was heated by coolant with 600°C from the inside and cooled with moderator water of 280°C from the outside, simulating conditions at the outlet of a hot assembly of the 2nd superheater. Stresses and deformations were predicted with ANSYS using homogenized anisotropic elements to simplify the honeycomb structure. A stiffening effect of the fuel rods has not been taken into account. Results are shown in Fig. 2.44. The maximum wall deflection towards the fuel rods is 0.3 mm. The peak stress in the corner pieces (zoomed in the right hand figure) is 233 MPa (equivalent total stress) which is less than the fatigue limit of around 300 MPa at 600°C.

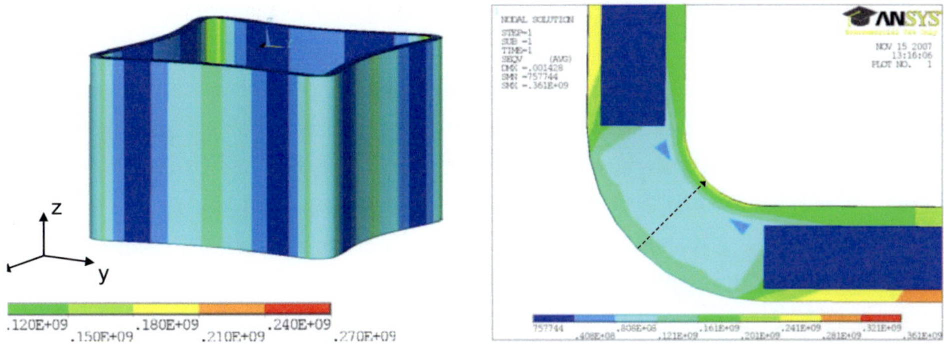

Fig. 2.44: Deformations (25x magnified) and stresses in Pa in a 2nd superheater assembly box [40]

Coolant temperature differences like those shown in Fig. 2.26 will also cause bending of the assembly box over the entire length. Assuming a radial power form factor of 1.21 and thus radial coolant temperature differences of +/- 40°C inside an assembly of the 2nd superheater, which is rather a theoretical worst case, we get thermal deformations of the assembly box as shown in Fig. 2.45. The maximum box deflection at half core height reaches almost 4 mm.

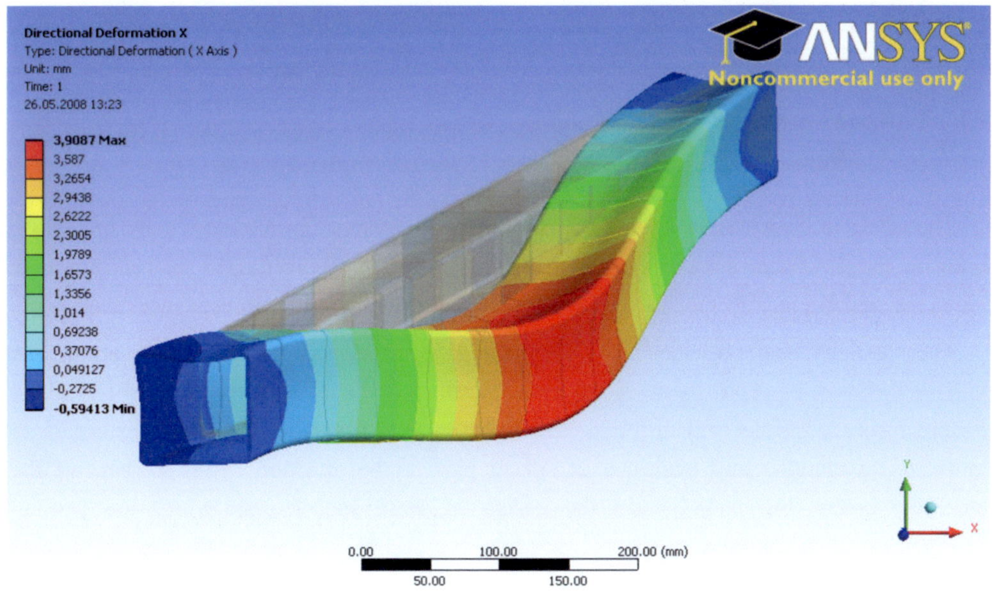

Fig. 2.45: Deformations in mm of an assembly box exposed to +/- 40°C radial coolant temperature gradients [48]

The effect of a change of the water gap between assembly boxes, due to such deflections, on the assembly power distribution has been studied with a 2D neutronic Monte Carlo analysis. An evaporator assembly cluster with 9 assemblies has been modified such that the inner assembly was displaced by 4.5 mm in x- and y-direction while the other assemblies were staying in place. Fig. 2.46 shows a comparison of the power peaking factor with fresh fuel without (left) and with a displacement, for comparison. The fuel has assumed to be fresh UO_2 with 6% enrichment and 2.5% Gd was added to the corner rods for excess reactivity compensation, causing low power there. Reflecting boundary conditions were assumed around the cluster. With regular assemblies, the local peaking factor is less than 1.2. As a consequence of the enlarged water gap, however, the local peaking factor increases to almost 1.4.

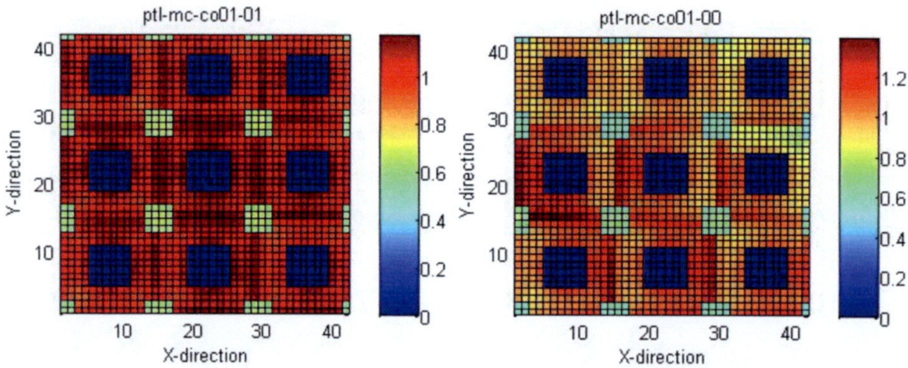

Fig. 2.46: 2D MCNP5 analysis of the relative power distribution of a regular assembly cluster (left) and with an inner assembly displaced by 4.5 mm each (right); fresh fuel [47]

The effect of a displaced assembly box becomes even more pronounced with increasing burn-up. Fig. 2.47 shows the same cluster after a burn-up of 20 GWd/tHM has been reached. Now, the peaking factor of a cluster with displaced central assembly exceeds more than 1.5.

These deflections are obviously not acceptable, and spacer pads between the assembly box will be required to avoid them. A systematical study with different axial power profiles, different core positions and different radial temperature gradients has been performed to optimize the number and positions of spacer pads to be attached to the corners of the assembly boxes. Fig. 2.48 shows the design of these spacer pads which are just an enlargement of the massive corner pieces of the boxes. Fig. 2.49 shows the bending line of the assembly box assuming 80°C temperature difference between the hotter and the colder side and 3 different axial power profiles which were assumed exemplarily. Two sets of spacer pads at 1.5 and at 3 m axial height are keeping the maximum deflection below 0.5 mm, which is small compared with the total gap width of 9 mm [41].

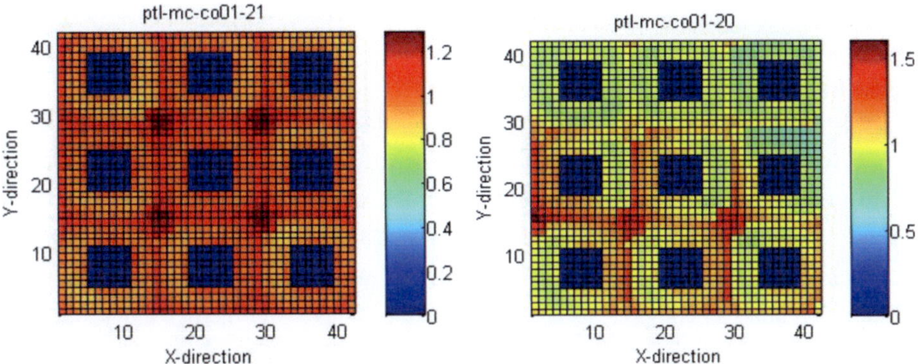

Fig. 2.47: Relative power distribution of a regular assembly cluster (left) and with an inner assembly displaced by 4.5 mm each (right); Burn-up 20 GWd/t_{HM} [47]

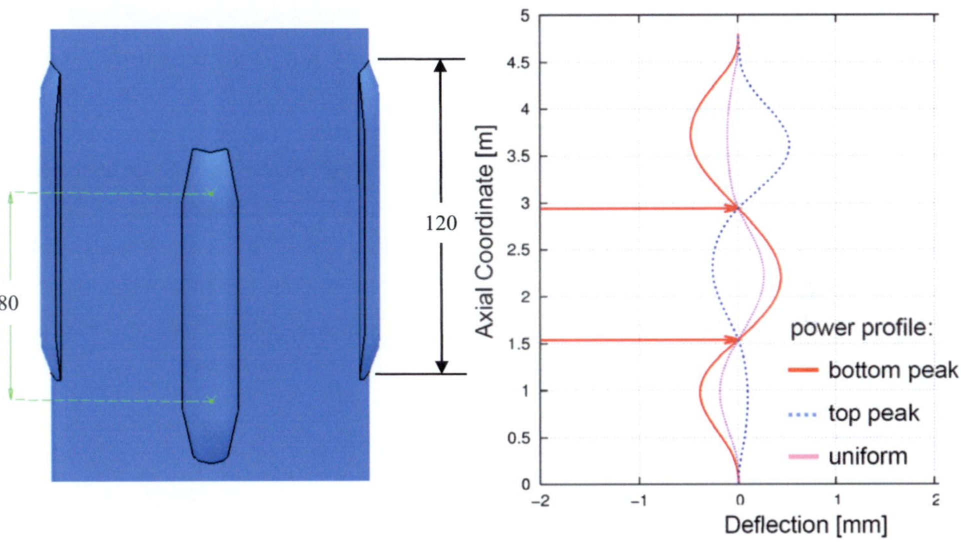

Fig. 2.48: Spacer pads of the assembly box (left) and bending line of the assembly box under worst case assumptions (right) [41]

Comparing the regular case with the bended case, Figs. 2.46 and 2.47, we got an increase of the peaking factor by around 20%. As only 1/9 of this deflection is to be expected at maximum, thanks to the spacer pads, we have to count for an unavoidable, statistical uncertainty of the local fuel rod power of around 2%.

2.10 Sealing against by-pass flows

Another consequence of the larger pressure difference will be the risk of ingress of moderator water into the coolant flow path. This will lower the coolant temperature locally, which means that it will increase in other areas accordingly, if the average core outlet temperature and the core power are kept constant. Therefore, the number of potential leaks has been minimized by welding all components as far as possible. Where ever thermal expansion must be allowed, sealing devices are foreseen to avoid leakage of moderator water into the mixing chambers and into head or foot pieces of the clusters.

The head pieces are sliding in the round openings of the steam plenum, which are sealed with C-rings as indicated in Fig. 2.4. Hofmeister et al. [13] show that these C-rings must be open to the high pressure side, i.e. the colder moderator side, to achieve best sealing performance. The openings of the steam plenum, Figs. 2.8 and 2.9, may deform, however, due to thermal expansion of the hotter core outlet chamber, which has been checked by Fischer et al. [42] with a finite element analysis. Fig. 2.49 shows the thermal expansion of 1/8 of the steam plenum, indicating that the outer diameter increases by up to 9mm at maximum core outlet temperature. A check of all openings by Redon [43] confirmed that the maximum excentric deformation from the ideal circular shape is less than 0.3mm, which is tolerable for a C-ring of 4 mm thickness to remain leak tight. Similarly, the excentricity of the 4 openings of the steam plenum at each extractable steam line, Fig. 2.8, is predicted to be less than 0.5mm, so that C-rings of 20 mm thickness may be applicable there as well.

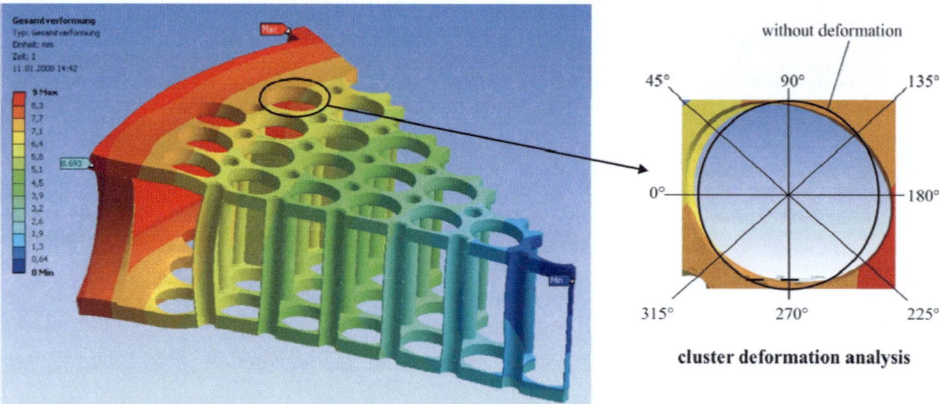

Fig. 2.49: Steady state thermal expansion of the steam plenum in mm and deformation of openings for assembly head pieces [43]

Each assembly box except the central one is sliding with a cylindrical extension in the bottom plate of the head piece, as indicated in Fig. 2.5. Two piston rings around these cylindrical extensions avoid leakage of colder moderator water of the gap volume into the coolant flow path inside assemblies.

The foot piece of each assembly is standing in round openings of the core support plate. Two piston rings as indicated in Fig. 2.6 are sealing the foot piece against leakage of colder moderator water of the gap volume into the lower mixing chamber.

The foot piece itself is made from two separate pieces which are clamped together with 8 bolts to give access to the fuel rod bundles for inspection and repair, as indicated in Fig. 2.6. Sealing at this joint is provided by a solid sealing lip of the diffuser piece. A similar sealing technique is foreseen at the joint between the central assembly box and the upper plate of the foot piece, indicated in Fig. 2.6, which are clamped together by 4 bolts. Schneider [44] predicts size and pre-stressing of these bolts to achieve leak tightness.

Finally, sealing devices are needed at the cylindrical extensions of the moderator boxes which are sliding in the insert of the foot piece, as shown in Fig. 2.6. Again, two piston rings are foreseen around each extension to avoid leakage.

The bypass flows through all residual gaps at these joints have not been assessed yet. They should be checked carefully in a later design stage with support from experiments. All analyses of this project were assuming that these sealings will remain perfectly leak tight.

2.11 Uncertainties of the analyses

A number of code uncertainties are contributing additionally to the hot channel predictions. Some of them were studied systematically, which will be summarized next.

The uncertainty of the sub-channel code C3CLM has been assessed for an unheated case by comparison with CFD results of Kiss et al. [29] for 1 axial wire lead, as illustrated in Fig. 2.31. The calculated cross flows through the gaps between fuel rods or between fuel rod and box walls are of special interest in this comparison, as they are responsible for the mixing behaviour inside the bundle. Two measures with respect to the C3CLM input were found to be necessary to bring the results of CFD and C3CLM closer together:

- Adaptation of the geometry of the corner channels to the data used in the CFD calculations

- Increase of the cross flow efficiency from 0.68 to 1.0.

Fig. 2.50 shows the results of the comparison with respect to the mass flow in the channels (averaged over 1 wire lead). There are 3 classes of mass flows:

≈ 0.026 kg/s for the outer corner channels

≈ 0.045 kg/s for the wall and inner corner channels

≈ 0.07 kg/s for the inner channels

A good agreement between CFD and C3CLM leaves a residual statistical uncertainty of 5.3%.

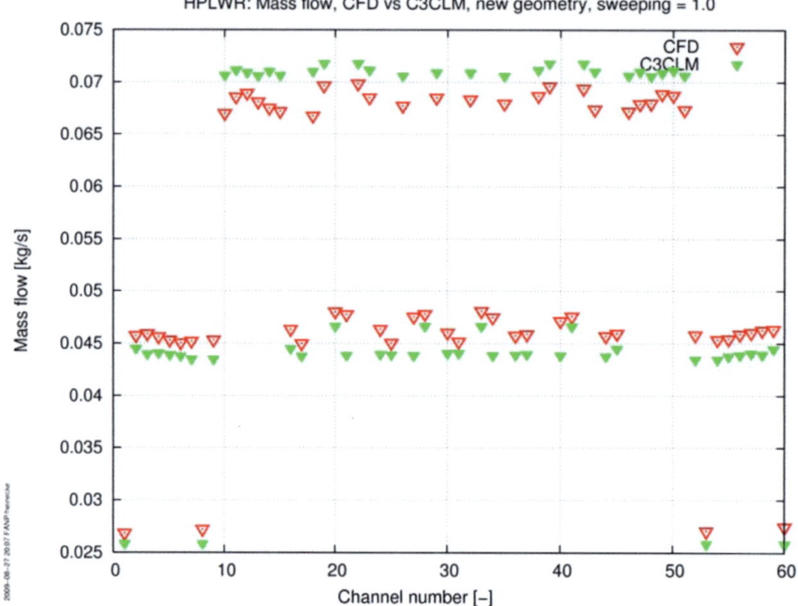

Fig. 2.50: Comparison of cross flow predictions with C3CLM sub-channel analyses and CFD analyses of Kiss et al. [29]

In order to study the influence of heating on the distribution of mass flows and temperatures, a comparison of C3CLM and MATRA, described by Kim et al. [45], has been performed for a heated bundle. Both codes use the power input from the coupled neutronics/thermal hydraulic analyses of Monti [21]. The statistics of the comparison is presented in Fig. 2.51. In total, 5100 calculated sub-channel mass flow values of a superheater assembly with the largest radial power gradient are taken into account. An average of 1.0016 and a standard deviation of 2.5% proof the good agreement between C3CLM and MATRA.

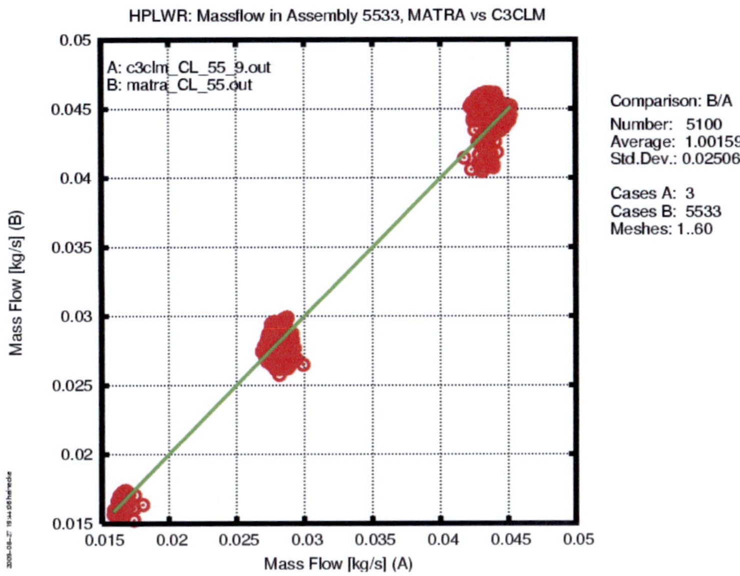

Fig. 2.51: Uncertainty of C3CLM mass flow predictions if compared with MATRA predictions [49]

Larger uncertainties of the thermal-hydraulic predictions arise from the heat transfer correlation, in particular in evaporator regions of high power density. These uncertainties have been evaluated by Loewenberg et al. [46] for different heat transfer correlations. The best correlation turned out to be the correlation of Jackson and Hall [18], which still has an uncertainty of 24%. The use of a look up table for prediction of heat transfer could reduce this uncertainty to around 10%.

The accuracy of the KARATE code system has been assessed by Maraczy et al [14] by comparing with test cases of boiling water reactors. The local power density of this code differed from the test cases by up to 1.4%. Similar studies were performed by Monti [21] by comparing ERANOS predictions with MCNP5 analyses for selected assemblies. The local power predicted by both methods differs by less than 5% in the high power region of the evaporator and by up to 11% in the superheater assemblies as shown in Fig. 2.52.

Further uncertainties in power predictions arise from the neutron scattering cross sections.

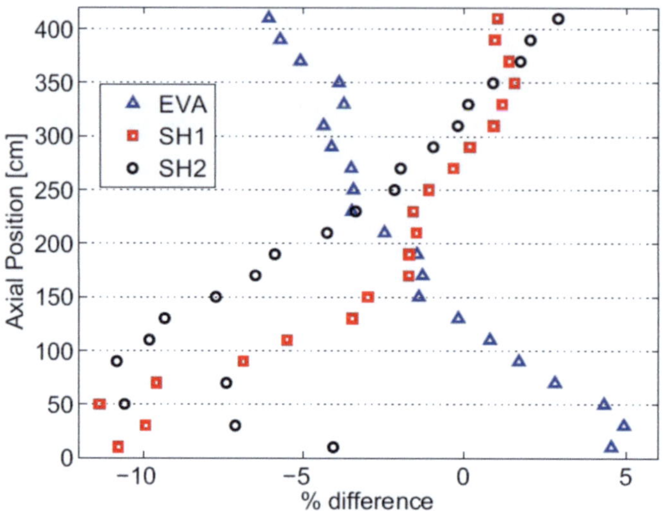

Fig. 2.52: Differences of local power predictions with ERANOS and MCNP5 for selected assemblies [21]

2.12 Hot channel assessment

The analyses summarized above lead to the following conclusions for the peak coolant outlet temperature of the hottest sub-channel of this core design.

In a first step, we derive the hot channel factors for coolant enthalpies. The radial peaking factors of assembly averaged coolant enthalpies throughout the burn-up cycle are a consequence of the radial power form factors, Figs. 2.20 and 2.21, at beginning and end of cycle (BOC and EOC, resp.), divided by the coolant mass flow rate of each assembly, Figs. 2.22 and 2.23. They range between 1.15 and 2.36 as listed in Tab. 2.11.

The local enthalpy peaking factors inside fuel assemblies are caused

- by the gradient of the neutron flux causing power peaking factors of individual fuel rods as shown in Fig. 2.25,
- by control rods, as illustrated in Fig. 2.28, and
- by Gd-poisoning of some fuel rods for compensation of excess reactivity, as illustrated in Fig. 2.29.

While the local power peaking factors exceed even a factor of 1.3, most of these non-uniformities are mixed out in the coolant by the wire wrapped around the fuel rods as shown in Fig. 2.30. As a conclusion, we need to account for a local peaking factor of the coolant enthalpy of 1.15 only.

Uncertainties arise primarily

- from bending of assembly boxes as shown in Figs. 2.46 and 2.47, of which effect only 1/9 has to be accounted here because of the spacer pads of these boxes,
- from uncertainties of neutronic and sub-channel codes, and
- from local blockage of the coolant flow path.

We can assume that these uncertainties are statistical errors, so that they sum up rather as the sum of variances. In total, however, an uncertainty of 10% is not considered to be too conservative.

		BOC			EOC		
Enthalpy peaking factors		**EVA**	**SH1**	**SH2**	**EVA**	**SH1**	**SH2**
Radial peaking factor		1.24	1.58	2.36	1.15	1.69	1.83
Local peaking inside FA		1.15	1.15	1.15	1.15	1.15	1.15
Uncertainties		1.10	1.10	1.10	1.10	1.10	1.10
Allowances		1.15	1.15	1.15	1.15	1.15	1.15
Total		**1.81**	**2.30**	**3.44**	**1.68**	**2.46**	**2.66**
Inlet mixing non-uniformity	kJ/kg	0	45	33	0	45	33
Average inlet enthalpy	kJ/kg	1417	2627	3075	1411	2503	3036
Peak inlet enthalpy	kJ/kg	1417	2672	3108	1411	2548	3069
Average outlet enthalpy	kJ/kg	2627	3075	3173	2503	3036	3173
Peak outlet enthalpy	kJ/kg	3603	3703	3445	3242	3859	3433
Peak outlet temperature	**°C**	638	673	584	521	731	580

Tab. 2.11: Hot channel assessment of peak coolant outlet temperatures for each heat up step

Finally, we need to account for allowance for operation and for the limited accuracy of the core and plant instrumentation. We assume a factor of 1.15 as a realistic guess.

If we multiply these coolant enthalpy peaking factors for each heat up step at BOC and EOC, we get the total peaking factors as shown in Tab. 2.11. They range between 1.81 and 3.44 at BOC, decreasing to 1.68 and 2.66 at EOC. In superheaters (SH1, SH2), these peaking factors exceed the target hot channel factor of 2 mentioned in chapter 2.2, whereas the peaking factors in the evaporator (EVA) have obviously some margin with this respect.

This result suggests to increase the power in the evaporator and to decrease it in the superheaters with respect to the envisaged power ratio of 4:2:1 (i.e. EVA 57%, SH1 29%, SH2 14%). The core design concept described here is following this strategy already to some extent. Figs. 2.15 and 2.16 shows a power split of 69%, 26% and 6% at BOC and 62%, 30% and 8% at EOC, for EVA, SH1 and SH2, respectively.

The average coolant enthalpies at the inlet and outlet of each heat-up step, Tab. 2.11, are a consequence of this power split. Due to the residual mixing non-uniformity of the upper and lower mixing chambers, Figs. 2.34 and 2.36, the peak inlet enthalpy is slightly higher by up to 45 kJ/kg at SH1 inlet and up to 33 kJ/kg at SH2 inlet. From these data, the peak coolant enthalpies at the outlet of each heat up step can be estimated as the peak inlet enthalpies plus the total peaking factor times the average enthalpy difference.

Finally, the steam table yields the peak coolant outlet temperature for each peak outlet enthalpy as indicated in Tab. 2.11. We get peak outlet temperatures beyond 600°C, which have obviously no more margin for the peak cladding temperature to stay below the material limits, in EVA and SH1 at BOC, and in SH1 at EOC, whereas the second superheater is not a cause for concern. Therefore, some further core optimization will be required to remove the remaining hot channels using the margins left in the rest of the core. The present result, however, is not too far from this optimum.

The peak fuel temperature is expected in the evaporator, where we predict a maximum linear heat rate of 39 kW/m at BOC decreasing to 32.5 kW/m towards EOC.

2.13 Discussion

The peak cladding temperature of this design is obviously exceeding the target of 630°C, as mentioned in Chapter 2.1, as the evaporator and first superheater coolant temperatures are already exceeding this limit at BOC and the first superheater peak coolant temperature is exceeding it even by far at EOC. Typically, we need to account for about 20°C to 30°C peak cladding surface temperature in excess of the peak coolant temperature, as predicted by Monti [21] for fresh fuel.

Moreover, the discharge burn-up of 32.5 MWD/t_{HM}, listed in Tab. 2.6, does not meet with the target burn-up of 60 MWD/t_{HM} with a fuel enrichment of 3% to 7% as listed in Tab. 2.3.

This discrepancy is also expressed by the hot channel factors, which meet with the initial expectations of Schulenberg et al. [5] only in the evaporator. The main reason for high peaking factors and low burn-up is the large size of the fuel assembly cluster. While the cluster design is appropriate in the evaporator region, where it enables a low form factor, easy fuel shuffling during revisions and standard control rod drives, the cluster size extends over the whole width of most of the superheater regions each. Thus, fuel shuffling from outside to inside, flattening the power profile, is disabled within the superheaters due to the constraint of clustering. Moreover, a compensation of enthalpy peaks by higher coolant mass flow rates in local superheater regions with higher power is disabled as long as the large assembly clusters can only be equipped with a common inlet orifice. Another reason for the limited burn-up is the use of stainless steel which is more neutron absorbing than Zircalloy and a higher percentage of structural material than in conventional light water reactors. The lesson learned from this study is that the small size assemblies should rather be shuffled individually and the cluster concept should rather be given up.

The biggest uncertainties of this core design turned out to be caused by heat transfer predictions, in particular in evaporator region with high linear power, and by material properties of the stainless steel claddings. Some realistic fuel assembly tests will be needed to reduce these uncertainties to acceptable limits.

Nevertheless, the results are already quite encouraging. As the peak coolant temperatures are around 600°C, in average, it is rather a question of further core optimization than a cause for a completely different core design concept to stay within the envisaged peak temperatures.

2.14 References

[1]   Y. Oka, S. Koshizuka, Y. Ishiwatari, A. Yamaji, Super Light Water Reactors and Super Fast Reactors, Springer, ISBN 978-1-4419-6034-4, 2010

[2]   M. Brandauer, M. Schlagenhaufer, T. Schulenberg, Steam cycle optimization for the HPLWR, 4th Int. Symposium on SCWR, March 8-11, 2009 Heidelberg, Germany, Paper 36

[3]   K. Dobashi, Y. Oka, S. Koshizuka, Conceptual design of a high temperature power reactor cooled and moderated by supercritical light water, Ann. Nucl. Energy 25, 487-505, 1998

[4]   M. Schlagenhaufer, J. Starflinger, T. Schulenberg, Plant control of the High Performance Light Water Reactor, Proc. GLOBAL 2009, Paris, France, Sept. 6-11, 2009, Paper 9339

[5]   T. Schulenberg, J. Starflinger, J. Heinecke, Three pass core design proposal for a high performance light water reactor, Progress in Nuclear Energy 50 (2008) 526-531

[6] J. Hofmeister, C. Waata, J. Starflinger, T. Schulenberg, E. Laurien, Fuel assembly design study for a reactor with supercritical water, Nucl. Eng. Design 237, pp. 1513-15121, 2007

[7] K. Fischer, T. Schneider, T. Redon, T. Schulenberg, J. Starflinger, Mechanical design of core components for a high performance light water reactor with a three pass core, Proc. GLOBAL 07, Boise, Idaho, USA Sept. 9-13, 2007

[8] H. Herbell, S. Himmel, Strukturmechanische Auslegung eines HPLWR Brennelement-kastens in Leichtbauweise, Forschungszentrum Karlsruhe, FZKA 7404, 2008

[9] M. Schlagenhaufer, B. Vogt, T. Schulenberg, Reactivity control mechanisms for a HPLWR fuel assembly, Proc. GLOBAL 07, Boise, Idaho, USA, Sep. 9-13, 2007

[10] C. Koehly, T. Schulenberg, J. Starflinger, Design concept of the HPLWR moderator flow path. 2009 International Congress on Advances in Nuclear Power Plants (ICAPP '09), Tokyo, Japan, May 10-14, 2009 Paper 9187

[11] A. Wank, Fluid dynamic design of complex mixing chambers, Dissertation University of Karlsruhe, Forschungszentrum Karlsruhe, FZKA 7520, 2009

[12] C. Waata, T. Schulenberg, X. Cheng, Coupling of MCNP with a Sub-Channel Code for Analysis of a HPLWR Fuel Assembly, The 11th International Topical Meeting on Nuclear Reactor Thermal-Hydraulics (NURETH-11), Avignon, France, October 2-6, 2005

[13] J. Hofmeister, T. Schulenberg, J. Starflinger, Optimization of a fuel assembly for a HPLWR, Paper 5077, Proc. ICAPP 05, Seoul, Korea, May 15-19, 2005

[14] C. Maraczy, G. Hegyi, G. Hordosy, E. Temesvari, C. Hegedos, A. Molnar, High Performance Light Water Reactor Core Design Studies, Proc. PBNC Conf. Aomori, Japan, 2008, Paper P16P1221.

[15] Cs. Hegedűs, Gy Hegyi, G. Hordósy, A. Keresztúri, M. Makai, Cs. Maráczy, F. Telbisz, E. Temesvári, P. Vértes, The KARATE Program System, PHYSOR 2002, Seoul, Korea, October 7-10, 2002

[16] A. Yamaji, Y. Oka, S. Koshizuka, Three-dimensional Core Design of SCLWR-H with Neutronic and Thermal-hydraulic Coupling, GLOBAL2003, New Orleans, USA, Nov. 2003

[17] M.J. Watt, C.T. Chou, Mixed Convection Heat Transfer to Supercritical Pressure Water, Proc. 7th Int. Heat Transfer. Conf. Munich, Vol. 3, 495-500, 1982

[18] J.D. Jackson, W.B. Hall, Forced convection heat transfer to fluids at supercritical pressure, in: Turbulent Forced Convection in Channels and Bundles, edited by S. Kakac and D. B. Spalding, Vol.2, Hemisphere Pub., 563, 1979

[19] K. Rehme, Pressure drop correlations for fuel element spacers, Nucl. Techn., Vol. 17, p. 15-23, 1973

[20] Cs. Maráczy, Gy. Hegyi, G. Hordósy, E. Temesvári, HPLWR equilibrium core design with the KARATA code system, 5th Int. Symp. SCWR, Vancouver, Canada, March 13-16, 2011, Paper 062

[21] L. Monti, Multi-scale, coupled reactor physics / thermal-hydraulic system and applications to the HPLWR 3 pass core, Dissertation University of Karlsruhe, FZKA 7521, 2009

[22] G. Rimpault, D. Plisson, J. Tommasi, J.M. Rieunier, R. Jacqmin, The ERANOS code and data system for fast reactor neutronic analyses. In: International Conference on the Physics of Reactors (PHYSOR), Seoul, Korea, October 7-10, 2002

[23] C. Murray, J. Staudenmeier, TRACE V5.0 User's Manual. Division of Risk Assessment and Special Projects Office of Nuclear Regulatory Research U. S. Nuclear, Regulatory Commission, Washington, DC 20555-0001, 2007

[24] W. S. Kim, Y.G. Kim, Y.J. Kim, A subchannel analysis code MATRA-LMR for wire wrapped LMR subassembly. Annals of Nuclear Energy, 29:303–321, 2002

[25] S.R. Himmel, A.G. Class, E. Laurien, T. Schulenberg, Sub-channel Analysis of a HPLWR Fuel Assembly with STAR-CD, 16th Pacific Basin Nuclear Conference (16PBNC), Aomori, Japan, Oct. 13-18, 2008, Paper P16P1152

[26] R. Nabbi, W. Bernnat, Application of Coupled Monte-Carlo and Burn-up Calculations for Detailed Neutronic Analysis for the FRJ-2 Research Reactor on High Performance Computers. ANS Topical Meeting in Monte Carlo, Chattanooga, TN, 2005

[27] S.R. Himmel, A.G. Class, E. Laurien, T. Schulenberg, Flow in a HPLWR Fuel Assembly With Wire Wrap Spacers, Proc. ANS/ENS Winter Meeting, Washington D.C., USA, Nov. 11-15, 2007

[28] S.R. Himmel, A.G. Class, E. Laurien, T. Schulenberg, Determination of Mixing Coefficients in a Wire-Wrapped HPLWR Fuel Assembly using CFD, Proc. International Congress on Advances in Nuclear Power Plants (ICAPP '08), Anaheim, USA, June 8-12, 2008

[29] A. Kiss, E. Laurien, A. Aszódi, Y. Zhu, Improved numerical simulation of a HPLWR fuel assembly flow with wrapped wire spacers, 4th International Symposium on Supercritical Water-Cooled Reactors, March 8-11, 2009, Heidelberg, Germany, Paper No. 03

[30] H. Pütz, Das Programm C3CLM, INTERATOM-Version des Programms COBRA IIIC INTERATOM, 33.02786.4, (1982)

[31] D.S. Rowe, COBRA IIIC: A digital computer program for steady state and transient thermal-hydraulic analysis of rod bundle nuclear fuel elements, Battelle, BNWL-1695, (1973)

[32] A. Wank, T. Schulenberg, E. Laurien, Mixing of cooling water at supercritical pressures in the HPLWR three pass core, NURETH-13, Kanazawa City, Japan, Sept. 27 – Oct. 2, 2009, Paper N13P1003

[33] A. Wank, Fluid dynamic design of complex mixing chambers, Dissertation University of Karlsruhe, Forschungszentrum Karlsruhe, FZKA 7520, 2009

[34] M. Möbius, A. Wank, T. Schulenberg, CFD Analysis of mixing in the headpieces of the High Performance Light Water Reactor, Diploma Thesis, Forschungszentrum Karlsruhe, IKET, Germany, 2008

[35] C. Kunik, B. Vogt, T. Schulenberg, Flow phenomena in the gap volume between assembly boxes, 4th International Symposium on Supercritical Water-Cooled Reactors, March 8-11, 2009, Heidelberg, Germany, Paper No.04

[36] C. Kunik, A. Miotto, T. Schulenberg, CFD-analysis of the moderator water flow in the HPLWR, Proc. ICAPP 10, San Diego, CA, USA, June 13-17, 2010, Paper 10055

[37] T. Ortega Gomez, Stability Analysis of the High Performance Light Water Reactor, Dissertation University of Karlsruhe, FZKA 7432, 2008

[38] X. Tiret, A. Wank, T. Schulenberg, CFD Analysis of the Decay Heat Removal in a Three-Pass Core with Downward Flow, Diploma Thesis, Institute for Nuclear and Energy Technologies, Forschungszentrum Karlsruhe, Germany, 2009

[39] T. Reiss, S. Fehér, Sz. Czifus, Calculation of xenon-oscillations in the HPLWR, Proc. ICAPP 09, Tokyo, Japan, May 10-14, 2009, Paper 9067

[40] H. Herbell, S. Himmel, T. Schulenberg, Mechanical Analysis of an Assembly Box with Honeycomb structure, 16th Pacific Basin Nuclear Conference (16PBNC), Aomori, Japan, Oct. 13 -18, 2008, PaperID P16P1150

[41] T. Schulenberg, J. Starflinger, C. Maraczy, J. Heinecke, W. Bernnat, Recent analyses of the HPLWR three pass core. 2010 International Congress on Advances in Nuclear Power Plants (ICAPP '10), San Diego, Calif., June 13-17, 2010, Paper 10157

[42] K. Fischer, T. Redon, G. Millet, C. Koehly, T. Schulenberg. Thermo-Mechanical Stress and Deformation Analysis of a HPLWR Pressure Vessel and Steam Plenum. Proceedings of ICAPP '08 Anaheim, CA USA, June 8-12, 2008, Paper 8050

[43] T. Redon, Auslegung der oberen Mischkammer eines Reaktors mit überkritischen Dampfzuständen, Diplomarbeit University of Karlsruhe, Forschungszentrum Karlsruhe, Institute for Nuclear and Energy Technologies, May 2007

[44] T. Schneider, Design proposal for the HPLWR three pass core, Fuel Assembly and Core Components, Studienarbeit University of Karlsruhe, Forschungszentrum Karlsruhe, Institute for Nuclear and Energy Technologies, 2007

[45] W.S. Kim, Y.G. Kim, Y.J. Kim, A subchannel analysis code MATRA-LMR for wire wrapped LMR subassembly. Annals of Nuclear Energy, 29:303–321, 2002

[46] M.F. Loewenberg, E. Laurien, A. Class, T. Schulenberg, Supercritical water heat transfer in vertical tubes: A look-up table, Progress in Nuclear Energy 50 (2008) 532-538

[47] W. Bernnat, Comparison of UOX and MOX for supercritical LWR, The 5th Int. Sym. SCWR (ISSCWR-5), Vancouver, Canada, March 13-16, Paper 067, 2011

[48] T. Reiss, Bending of the HPLWR outer assembly box, IKET, Forschungszentrum Karlsruhe, unpublished, 2008

[49] J. Heinecke, Subchannel analysis review and comparison of codes, AREVA NP, unpublished, 2009

[50] J. Heinecke, Feasibility of the core concept, AREVA NP, unpublished, 2009

[51] T. Schulenberg, C. Maráczy, W. Bernnat, J. Starflinger, Assessment of the HPLWR core design, The 5th Int. Symp. SCWR (ISSCWR-5), Vancouver, Canada, March 13 – 16, Paper P057, 2011

[52] Gy. Hegyi, C. Maráczy, E. Temesvári, HPLWR equilibrium core design, KFKI Atomic Researach Institute, unpublished, 2009

[53] T. Reiss, S. Fehér, Sz. Czifrus, Xenon oscillations in SCWRs, Progress in Nuclear Energy 53, pp. 457-462, 2011

3   Primary System

The primary system of a pressurized water reactor comprises the reactor pressure vessel (RPV), the steam generators, recirculation pumps and the pressurizer. The primary system of a boiling water reactor is reduced to the reactor pressure vessel in which the recirculation pumps and the steam separators and dryers are located. The primary system of a supercritical water cooled reactor, however, reduces this design effort to the bare minimum: It is the reactor pressure vessel including the core, the core barrel, a mixing plenum each above and underneath the core, and the control rod drive. There is no more closed coolant loop in the primary system, neither inside nor outside the reactor which simplifies the design enormously. On the other hand, the higher temperature differences, a higher system pressure and an increased pressure drop are causing new challenges beyond the state of the art.

3.1   General design strategy

A system pressure of 25 MPa at the reactor inlet will require an enlarged wall thickness of the pressure vessel. In particular in combination with an increased core outlet temperature, this might cause larger thermal stresses which can be minimized by shielding the hot steam outlet from the thick walled structures. Moreover, the coolant heat up from 280°C to 500°C will cause larger difference of the thermal expansion of fuel assemblies, compared with a pressurized water reactor, which requires leak tight, sliding joints of the head pieces of the assembly clusters and at other locations. Finally, the larger pressure drop of the coolant from inlet to outlet requires tighter sealing concepts, and the components confining the coolant flow path should preferably be welded wherever possible.

One of the first conceptual designs for the vessel, published by Kataoka et al. [1] in 2003, was planned for an output of 950 MWe and featured outer dimensions similar to a pressurized water reactor. A first dimensioning of the shell suggested an inner diameter of about 4.3 m, a total height of 15 m and a wall thickness of 0.39 m. The inside of the reactor pressure vessel (RPV) wall was cooled by feed water from the inlet, while the hot steam plenum was covered with a thermal insulator to keep the supercritical steam separated from the surrounding coolant. A disadvantage of this design was the lack of insulation of the hot steam pipe, connecting the steam plenum with the outlets in the RPV. With such design, thermal stresses and deformations are likely to occur in the RPV, leading eventually to leakage through the vessel flange.

Buongiorno [2] and Buongiorno and MacDonald [3] used a similar approach for their RPV design, but for a higher electrical output of 1600 MWe. The shell was dimensioned based on the ASME Code; its structural performance was validated using three-dimensional finite-element analyses. Their analyses led to a height of 12.40 m, an inner diameter of 5.30 m and a maximum wall thickness of 0.46 m. In addition to the RPV inner wall, the hot steam pipe connection was cooled with feed water from outside, preventing contact between the hot pipes and the outlet nozzles. In both concepts, two inlets and two outlets were foreseen.

A third concept, introduced by Bittermann et al. [4], featured three combined inlets/outlets using a concentric pipe configuration. The inner pipe was connected to the hot steam plenum and released the superheated steam outside of the vessel. The outer pipe, which was the inlet nozzle in this case, was used to supply the core with feed water and additionally worked as a thermal sleeve to prevent contact between the outlet steam and the inner wall of the RPV. A spring at the end of the inner tube was proposed to compensate the thermal expansions. Furthermore, piston ring seals were provided to prevent mixing of the inlet and outlet mass flow. Such a strongly cooled steam pipe, however, reduces the steam temperature.

Starting from these preliminary considerations, the design of the pressure vessel and its internals of the HPLWR is based on the following general strategy:

The fuel assemblies or assembly clusters are standing on the core support plate forming the bottom of the core barrel which, in turn, is suspended in the large reactor flange. As a consequence, the assemblies will grow upwards the more they are heated up. The hot steam plenum, mixing the steam between the evaporator outlets and the 1st superheater inlets, as well as collecting the superheated steam at the 2nd superheater outlet, is resting on struts of the RPV close to the outlet flanges to keep the steam plenum in place, independent from the core expansion. This concept will ease sealing of the outlet steam pipe and thus minimize the bypass flow into the superheated steam. It requires, however, that the head pieces of the assembly clusters are sliding by several centimeters in the mixing plenum due to their thermal expansion.

The feedwater is supplied through separate inlet flanges in the upper half of the RPV, and the coaxial steam pipes are cooled from outside by feedwater, flowing around them with low velocities, to keep the pressure vessel cold, but also to minimize heat losses of the steam.

The control rod drives shall be mounted on top of the closure head like in a pressurized water reactor and control guide tubes in the upper half of the RPV shall be designed accordingly. The control rods shall run inside the moderator boxes to avoid cold feedwater leakage into the steam.

3.2 Design of the reactor pressure vessel and its internals

The dimensioning of the RPV and closure head as well as the design calculation for the studs, nuts and O-ring seals was performed using the safety standards of the nuclear safety standards commission (KTA) in Germany.

The proposed characteristics were modified by applying an increased safety factor of 115% on the design pressure. In this case, for an operating pressure of 25 MPa, a value of 28.75 MPa was used, giving a higher safety margin. For a design temperature of 350 °C, the tensile strength of the proposed vessel material 20 MnMoNi 5 5 steel depends on the thickness of the component as depicted in Fig. 3.1. For a wall thickness of 300 mm, the tensile strength is around 590 N/mm^2. Creep has not been considered, since the temperatures will not exceed 350°C at the inner wall of the RPV.

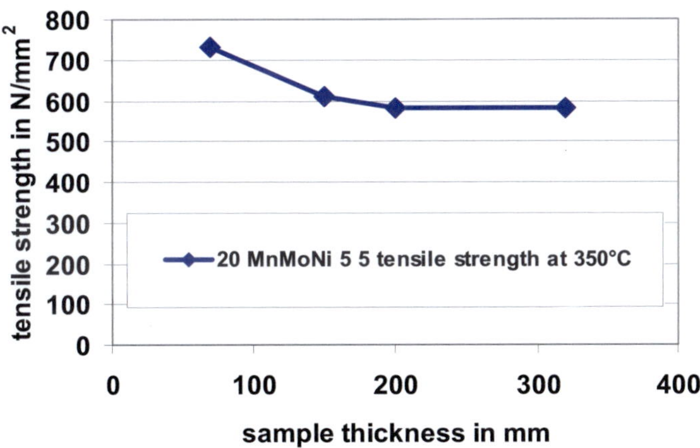

Fig. 3.1: Tensile strength of 20 MnMoNi 5 5 for a design temperature of 350°C and different sample thicknesses

The resulting characteristics of the RPV and its internals are listed in Tab. 3.1. Upper and lower part of the pressure vessel are connected with 40 bolts of size M210x8. The material of the steam outlet flange agrees with the steam line material P91, which can stand higher temperatures than the RPV material.

The reactor design is based on an earlier proposal by Fischer et al. [5] which was later updated by Koehly et al. [6]. Core components were designed by Fischer et al. [7] according to the design concept described in Chapter 2. Fig. 3.2 shows a segment of the RPV with outlets, the core barrel, reflector, control rod guide tubes and steam plenum. For simplification, only one inserted fuel assembly cluster per heat up step is displayed.

Parameters RPV	
Operating/design pressure	25.0/28.75 MPa
Operating/design temperature	280/350°C
Number of cold/hot nozzles	4/4
Dimensions RPV	[m]
Height (including closure head)	14.29
Height (excluding closure head)	11.42
Inner diameter	4.46
Wall thickness (cylindrical shell)	0.45
Wall thickness (bottom head)	0.30
Wall thickness (upper flange)	0.56
Wall thickness (closure head)	0.40
Material RPV/internals	
Vessel, closure head	20 MnMoNi 55 (SA 508)
Outlet flange, steam pipe	P91
Internals except core	1.4970
Weight RPV/internals	[t]
Lower vessel	534
Closure head with nuts and bolts	122
Internals except core	261

Tab. 3.1: Characteristics of the reactor pressure vessel and its internals

The core barrel is suspended in the reactor flange, i.e. at the lower vessel top and centered in radial direction using four centering logs. The core barrel sits, together with the control rod guide assembly, on a ledge machined from the RPV flange and is preloaded with a spring element. The lower vessel is bolted with the closure head flange using reduced shank bolts and nuts. Two O-ring seals ensure leak tightness between the closure head and the lower reactor pressure vessel.

For the three pass core design, the coolant has to be mixed in the mixing chambers above and underneath the core. The spherical lower mixing plenum, which is welded to the bottom of the lower core plate, homogenizes the water flow from the downcomer before it enters through the core support plate into the lower part of the evaporator. In the peripheral section of the lower mixing chamber, welded with the core support plat, the coolant from superheater 1 is mixed before it enters superheater 2.

The heated, supercritical steam is collected and mixed above all fuel assembly clusters in the steam plenum. The space above the steam plenum is designed to house the guide tubes for the control rods. Each guide tube is centered individually at the top of the steam plenum using the corresponding head piece bushing of the fuel assembly cluster proposed by Hofmeister et al. [8].

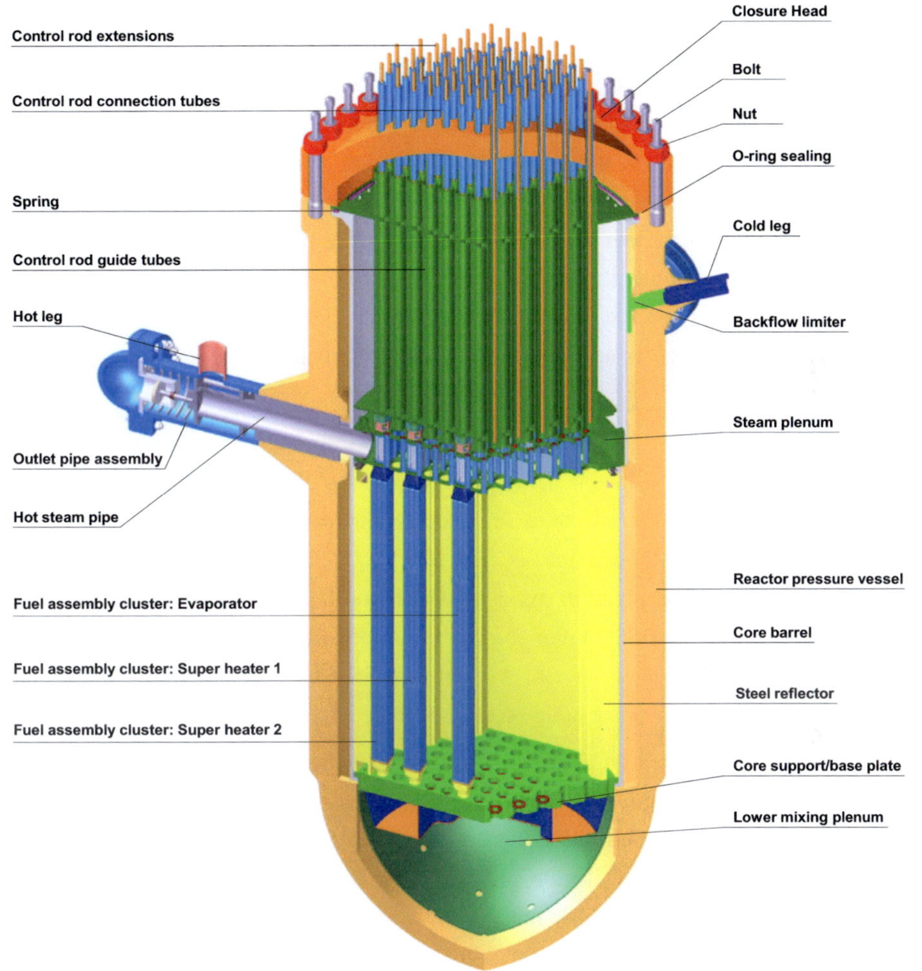

Fig. 3.2: Reactor pressure vessel and its internals [6]

The design challenges for the RPV are the thermal expansions between the core barrel, the steam plenum and the RPV. If the steam plenum would be allowed to move with the expanding core barrel during operation, the fixed hot pipes would jam inside the steam plenum and the C-ring seals would fail. Therefore, in this design, the steam plenum rests instead on the support brackets inside the lower vessel, while the core barrel is suspended at the closure head flange. With this concept, thermal expansions between the internals are decoupled and thermal stresses are minimized.

To remove the steam plenum, the control rod drives are disconnected from the spider, the closure head is opened and lifted, and the control rod guide tubes are taken out as an entire unit. Then the 4 steam outlet pipes are pulled back by hydraulic pistons in the steam outlet flanges. Now, the steam plenum can be taken out, giving access to each individual assembly cluster for shuffling or replacement.

The arrangement is assembled again in the opposite order: After completion of the core, the steam plenum is lowered into the pressure vessel, guided by rails of the core drum, and mounted over the head pieces of the assembly clusters until it sits on the support brackets of the pressure vessel. Now the hydraulic pistons are released and springs push the steam lines back into the steam plenum. Four fins on each side of the extractable steam pipes align the pipes to ensure that they run gently into the steam plenum. A C-ring around each steam pipe seals it against feedwater leakage into the steam plenum. Next, the control rod guide tubes are inserted, the closure head is added and bolted, and the control rod drives are connected with the spiders again.

Tight sealing against leakage of moderator water into the steam is an important design requirement of the HPLWR concept. Leakage would lower the steam temperature. As the core outlet temperature is controlled to an average outlet temperature of 500°C during operation, however, a cold streak means that hot spots will appear somewhere else to compensate the cold water leakage in average, which could even cause degradation of fuel rods. The sealing concept is challenging as coolant temperature differences are significantly higher with this concept, compared with conventional light water reactors, so that larger thermal expansions between reactor components are to be expected.

The steam plenum is heated up to around 500°C at its outer mixing chamber. The sealing concept against ingress of moderator water into the steam is solved there with two C-rings around each head piece of the fuel assembly clusters, shown in Fig. 2.4, and a C-ring around each steam line, as described by Hofmeister et al. [8], shown in Fig. 3.3. The downcomer water is sealed from the inter-assembly moderator water by a bellow which is pressed by the steam plenum onto the top of the reflector. This concept allows larger openings in the core barrel for the support brackets of the RPV to carry the steam plenum. Vertical connection tubes between top and bottom plate of the steam plenum are closed at the top plate. They serve as stiffeners of the steam plenum and as guide tubes for the core instrumentation.

The thermal deformations of the steam plenum at its openings for the head pieces of the assembly clusters and its influence on leakage are discussed in Chapter 2.10.

Sealing of the steam line inside the outlet flange is foreseen with piston rings around both disks which separate the hot steam outlet compartment from the purging feedwater flow around the steam line. The sealing concept of the foot pieces is described in Chapter 2.10 and shown in Fig. 2.6. The lower mixing plenum is a welded construction which can be regarded as leak tight.

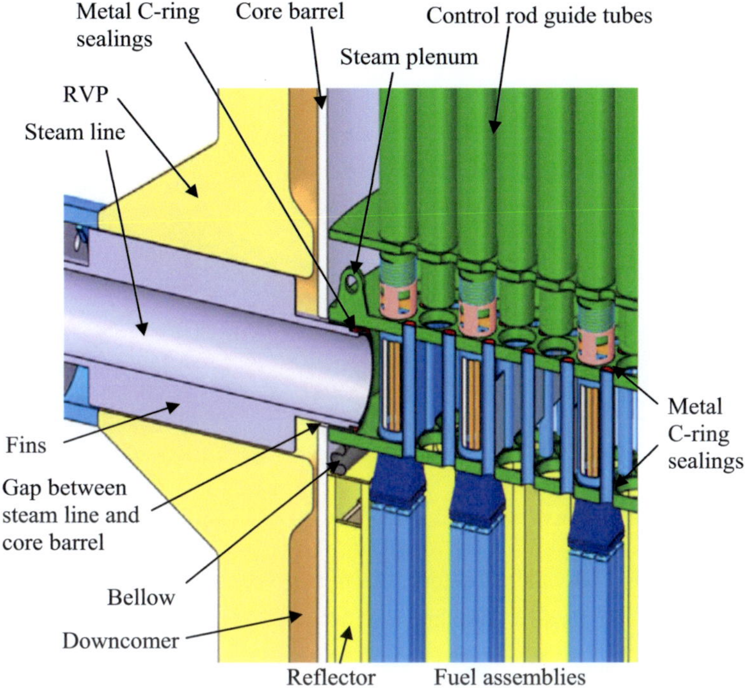

Fig. 3.3: Sealing concept of the steam plenum [6]

3.3 Analysis of the feedwater flowpath

3.3.1 Backflow limiter

The feedwater enters the RPV through 4 inlet flanges, which are equipped with a backflow limiter each to minimize coolant losses in case of a large break of a feedwater line. This passive safety device been optimized by Fischer et al. [10] to yield a minimum pressure loss in forward direction and a maximum losses in backward direction. The physical mechanism of such a backflow limiter is a vortex diode, which causes high circumferential velocities and flow separation in backwards direction. The design chosen here is explained with Fig. 3.4.

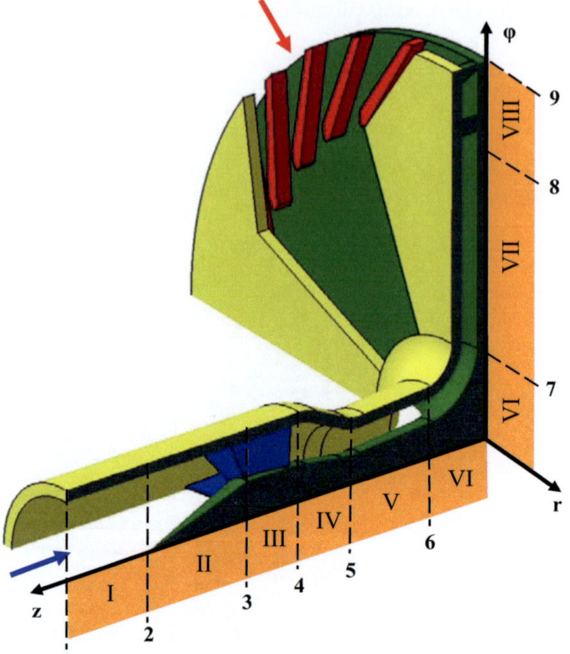

Fig. 3.4: Quarter section of the backflow limiter with indication of flow sections [10]

For regular operation, feedwater is entering in forward direction in section I. The 10 inlet swirler vanes in section III cause a swirl angle of around 10° to 15° to avoid flow separation in the radial diffuser VI to VII. More swirl should be avoided to minimize the pressure loss in forward direction, whereas less swirl will risk a flow separation. The 30 outlet swirler vanes are designed with a swirl angle of 60° which has only little effect on the pressure loss since the velocity is only around 7m/s at the swirler inlet. The proposed dimensions of the backflow limiter are listed in Tab. 3.2. The flow has been studied with CFD for forward and backward direction. The predicted pressure drop in forward direction is shown in Fig. 3.5 for different angles of the inlet swirler.

Position	1	2	3	4	5	6	7	8	9
Inner Diam.	0	0	0.1	0.1	0.08	0.09	0.14	0.37	0.48
Outer Diam.	0.2	0.2	0.2	0.2	0.14	0.16	0.14	0.37	0.48
z	0.7	0.59	0.45	0.38	0.3	0.2	0.02	0.02	0.02

Tab. 3.2: Dimensions of the backflow limiter in meters as proposed by Fischer et al. [10]

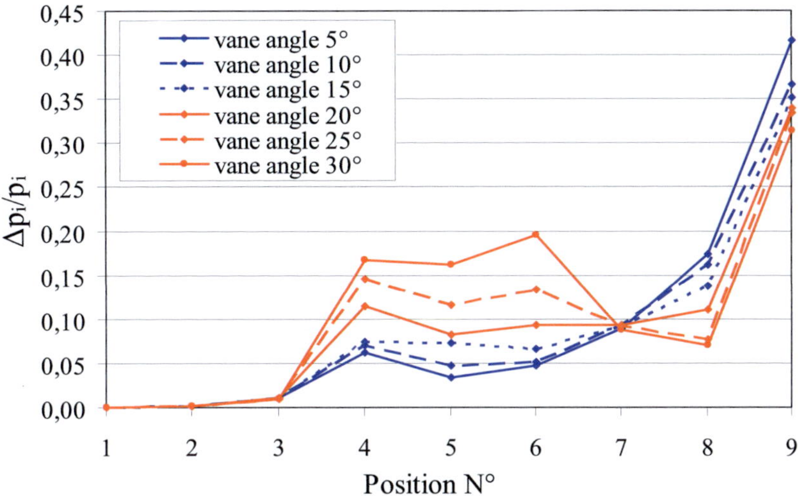

Fig. 3.5: Related pressure drop in forward direction for different swirl angles [10]

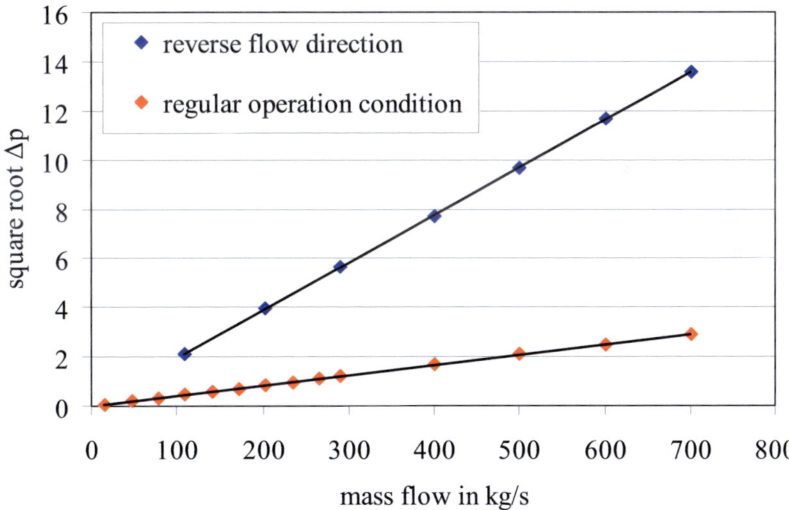

Fig. 3.6: Square root of the pressure loss in forward and backward direction [10]

In backward direction, the outlet swirler, which becomes the inlet swirler now, causes a high circumferential velocity which is further accelerated with decreasing diameter of the nozzle. Fischer et al. [10] predict 3 flow separation zones at position 5, between the inlet swirler vanes in section III and finally at the hub in section II. The resulting pressure loss for single phase flow in both directions is plotted in Fig. 3.6. At a given mass flow, the pressure drop in reverse direction is 21 to 22 times higher than in forward direction.

Beyond a mass flow of 700 kg/s, the pressure near the hub will be less than the saturation pressure under reverse flow conditions, even with full supercritical pressure inside the reactor, causing cavitation, followed by choking of the two-phase flow. The CFD analysis could not be extended to this range, but a further increase of mass flow can hardly be expected then.

3.3.2 Moderator flowpath

Once the feedwater has been entering the RPV, it is split in the HPLWR concept into a downward flow, line A, through the downcomer and from there to the core inlet, and an upward flow, line B, to the closure head and from there through the control rod guide tubes to the top of the core. The originally foreseen flow path of moderator water, with downward flow of inter-assembly gap water in the core, was causing a flow reversal of moderator water there. The consequence would have been an uncontrolled moderation of the core as shown by 2D analyses of Kunik et al. [11]. Even an increase of the gap water mass flow rate to 75% could not prevent the flow reversal and the associated risk of a neutronic feedback of moderator water on the core power distribution which is hardly predictable. This unstable flow structure was avoided later by the following optimized flow path, where the gap water flows upwards under any conditions.

The improved moderator flow path consists of a serial moderator flow path instead of a parallel one. The total mass flow rate is split after entering the reactor pressure vessel into 50% downcomer flow (line A) and 50% upward flow, which is used first as moderator inside the inner boxes (downward flow), then as moderator in the assembly gaps (upward flow) and finally as reflector water (downward flow) (line B). Because heat up of the downcomer water (line A) is negligible, this part is colder and both lines had to be mixed uniformly in the lower mixing plenum underneath the core before entering the evaporator assemblies (line C). The assumed mass flow rates are shown in Fig. 3.7. This new flow path was enabling stable upwards flow conditions of the gap water under any thermal load of the core.

The mass flow split is a function of pressure resistances in each line. It can be adjusted with orifices in lines A and B, dimensioned for the envisaged pressure drop of both lines. The steady state pressures and mass flow rates under full load conditions were determined with pressure drop correlations for pipes and orifices. Assuming that all control rods which are running inside the moderator boxes are withdrawn to their highest positions, a pressure drop of 67 kPa has been calculated for line B. The main pressure drop (33 kPa) in this line is caused by the inlet orifices of the moderator boxes above the core, Fig. 2.5 (left), where the moderator water is still at 280°C, to assure that the total flow resistance of these boxes is larger than the hydrostatic pressure differences in case of different heat up of moderator water. Orifices in the spherical lower mixing chamber underneath the core leading the downcomer flow into the mixing volume, Fig. 3.8, adjust the mass flow rate of line A. A different mass flow split between lines A and B, if needed, could easily be obtained by simply changing size or number of these orifices.

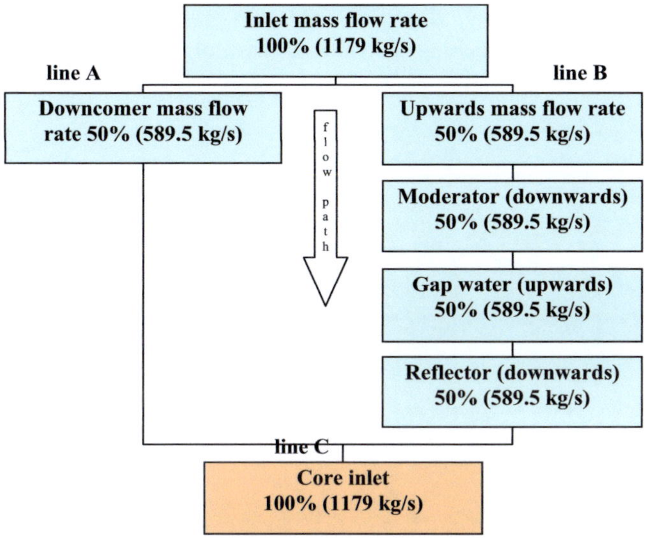

Fig. 3.7: Split of the feedwater mass flow into moderator and downcomer flow [6]

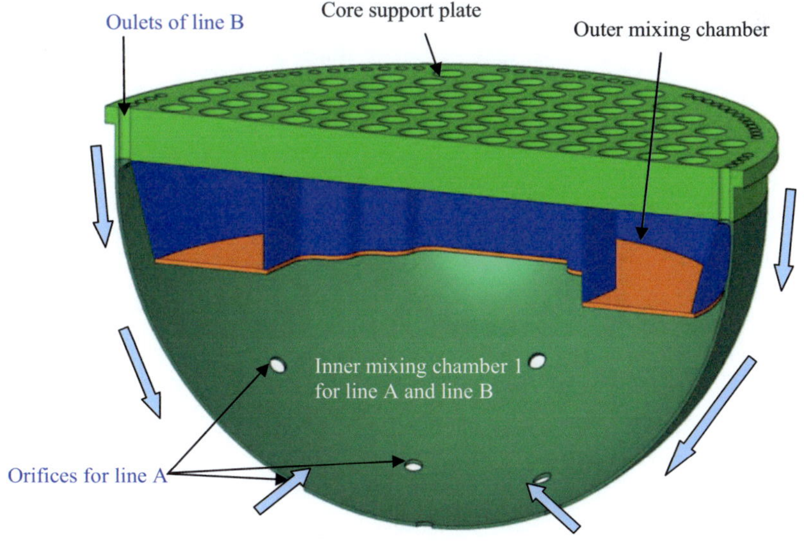

Fig. 3.8: Orifices in the lower mixing chamber adjusting the feedwater mass flow split [6]

The 50% mass flow fraction of line B is guided first from the inlet flange to the closure head to cool the reactor pressure vessel uniformly to feedwater temperature. Penetration slots in the suspension of the core barrel and of the top plate of the control rod guide tube arrangement at the reactor flange, shown in Fig. 3.9, have been dimensioned sufficiently large to minimize the pressure drop, but small enough that the remaining structure of the core barrel suspension can still carry the weight of the core. These slots cause a pressure drop of 680 Pa under nominal conditions. Openings between the control rod guide tubes, Fig. 3.10, as well as the open cross section inside the guide tubes provide a large total flow cross section for downward flow.

Fig. 3.9: Penetration slots in core barrel for upwards flow into the upper dome [6]

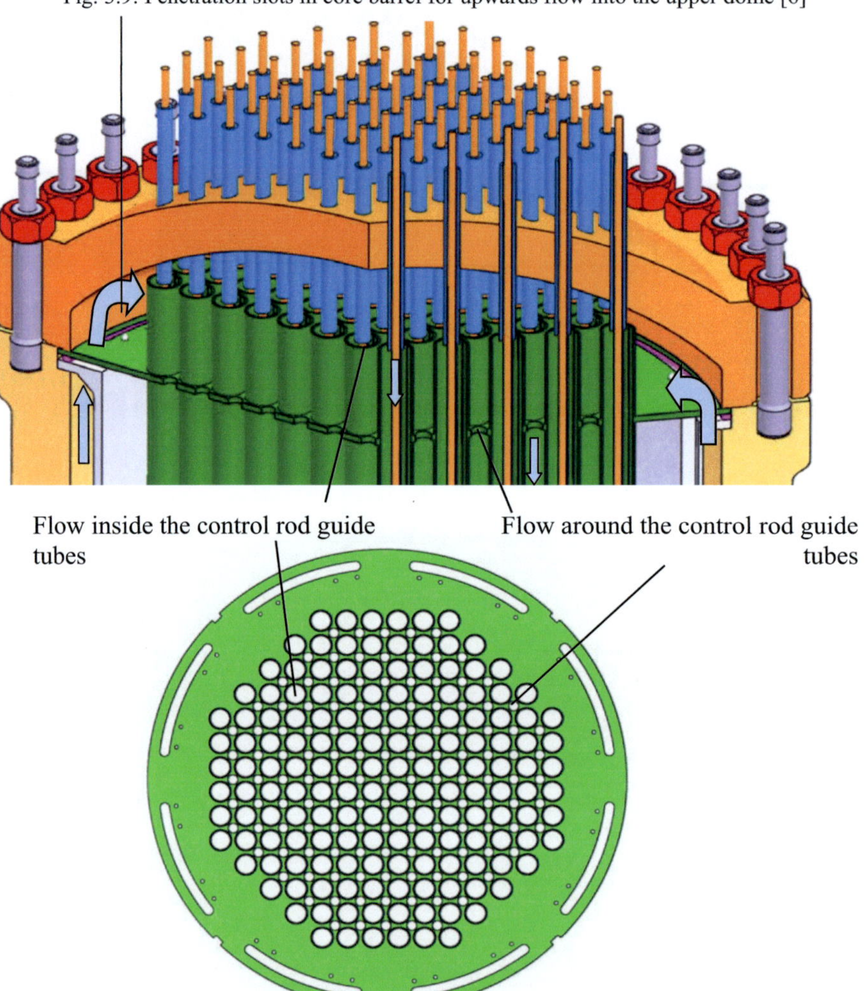

Fig. 3.10: Openings for feedwater flow through and around the control rod guide tubes [6]

At the upper side of the steam plenum, the moderator water enters the inlets of the moderator boxes inside the nine assemblies of each cluster. Five of these nine moderator boxes guide the control rods as indicated in Fig. 2.5. A square tube with 20.88 mm outer side length and 13.88 mm inner side length has been selected as control rod. They leave a gap of 1 mm between control rod and moderator box, which is purged with moderator water. A small opening of 4 mm diameter in the control rod spider provides an additional mass flow inside the control rod. The control rod spider has a coupling at the top to be connected with the control rod drive. As the head piece of the assembly cluster needs to be cylindrical to seal it at the penetration holes in the steam plenum, the four corner boxes need to be bent to fit into the same head piece. As a consequence, they cannot be equipped with control rods unless these are flexible. Round extensions of the square moderator boxes have been selected there to ease bending. An inlet orifice, shown with red in Fig. 2.5 (left), is dimensioned with an orifice diameter of 11 mm such that the same moderator mass flow rate will be obtained through all moderator boxes to get approximately the same heat up.

3.3.3 Downcomer flow

The downcomer flow, line A in Fig. 3.7, is needed to keep the RPV at feedwater temperature and thus also to minimize thermal stresses in this thick walled component. The flow structure has been analyzed by Foulon et al. [12] with the CFD software package STAR-CD version 3.26. Only steady state analyses have been performed and the standard high-Reynolds k-ε-model has been chosen. Due to symmetry, only ¼ of the total RPV has been modelled and the cut surfaces of the fluid domain were modelled as cyclic boundary conditions. The complex geometry of the back flow limiter has been simplified to obtain a mesh with reasonable complexity and a reasonable number of cells for the simulation. The mesh has been generated with ProAmm, a sub program of STAR-CD and consisted of 570000 hexahedral cells.

The inlet boundary, given by the backflow limiter described by Fischer et al. [10], was simplified as an annular surface. The inlet velocities were averaged from the detailed analysis, resulting in a radial component of 7.7 m/s and a circumferential component of 9.8 m/s. A steam temperature of 500°C was assumed inside the steam pipe, whereas the RPV was assumed to be thermally insulated. This early analysis was assuming a flow split of 25% upwards and 75% downwards.

Besides the outlet region of the backflow limiter, where peak velocities of the feedwater up to 12 m/s were reached, the flow in the downcomer was less than 1 m/s. In the region around the steam outlet pipe, the velocities were very small reaching values only between

0.03 m/s and 0.25 m/s and the flow did not show a uniform direction. Heat transfer coefficients in the downcomer and around the steam pipe, which were obtained from this analysis, are shown in Fig. 3.11. They served later as boundary conditions for transient stress analyses of the RPV.

The highest values of the heat transfer coefficients were obtained near the outlets of the backflow limiter, where jets collide with the wall of the RPV reaching values up to 50000 W/m²K as shown in Fig. 3.11 (left). Around the steam outlet pipe, the heat transfer coefficients of the RPV were significantly lower due to the very small velocity of the feedwater. At most of the surface of the core drum, the heat transfer coefficients were almost homogenous except near the steam outlet pipes, where they reached a maximum of 15000W/m²K due to the high velocities in the small gap between core drum and RPV.

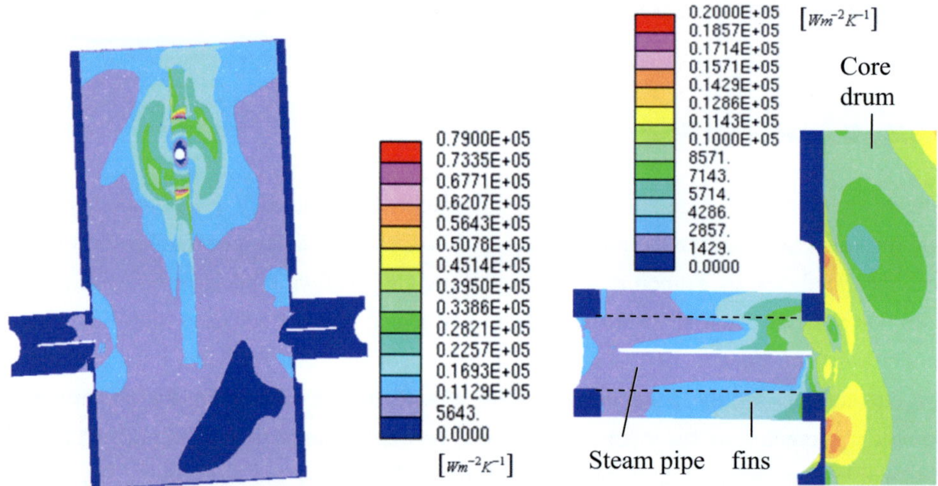

Fig. 3.11: Heat transfer coefficient of feedwater at the inner wall of the RPV (left) as well as on the core drum and on outer surface of the steam pipe (right) [12]

The coolant temperature distribution is shown in Fig. 3.12. As predicted, the highest temperatures (maximum 330°C) are in the region of the steam outlet pipe. For the rest of the analyzed volume, an almost homogenous temperature of around 280°C was obtained. A warm jet along the RPV surface with 5°C hotter temperature was found in the wake behind the steam outlet pipe, which is zoomed out in Fig. 3.12 (right).

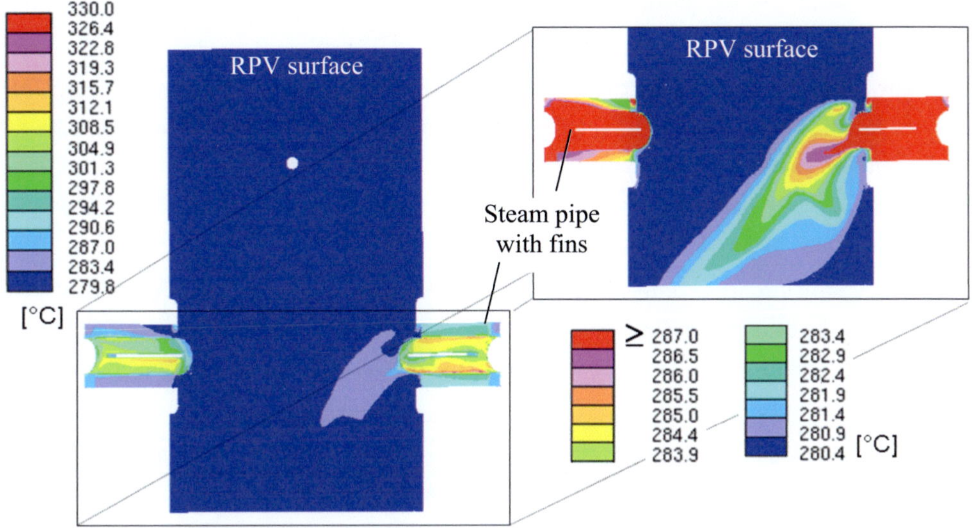

Fig. 3.12: Coolant temperature at the RPV and steam pipe surfaces [12]

3.4 Stress analyses

3.4.1 Reactor pressure vessel

These thermal boundary conditions were applied then in a finite element analysis with ANSYS by Reiss et al. [13] to predict transient temperatures and stresses of the RPV. Regions around the inlet and outlet flanges were of particular interest.

Two hypothetical transients were examined exemplarily as realistic transients had not yet been available at that time:

1. A case called "normal shut-down": starting from steady-state, part load conditions with a pressure of 28.75 MPa, a mass flow rate of 20% of the full power operating state and a temperature of the coolant outside the boundary layer between 280°C to 300°C as shown in Fig. 3.12. During the transient phase, the pressure and heat transfer coefficients are assumed to remain constant, while the temperature drops with 50 K/h in the first 2 hours and with 30 K/h in the next 3 hours, starting from 280°C.

2. Another case called "fast shut-down": the starting steady-state conditions are the same as before. The difference is that the initial mass flow rate was at 100%. During the transient,

the pressure drops from 28.75 MPa to 0.5 MPa and the temperature from 280°C to 80°C in 720 seconds each. The mass flow rate is reduced to 50% of the nominal value.

For all analyses, it was assumed that the heat transfer coefficient depends linearly on the mass flow rate. The heat transfer coefficient at full mass flow rate was given by the CFD analysis as shown in Fig. 3.11. On the outside surfaces of the RPV, the heat transfer coefficient was estimated around 10 W/m^2K and the ambient temperature was assumed to be 150°C there. These values remain constant during both cases. 20 MnMoNi 5 5 has been assumed as vessel material.

The RPV structure was cut for the ANSYS analysis into an upper part of the lower vessel, which yields the stresses around the inlet flange, a lower part around the outlet flange, and the thin-walled outer steam flange in separate analyses. Cut surfaces were assumed to remain plane during the deformations.

Linear elastic analyses were performed in all cases. Unfortunately, low cycle fatigue data of irradiated vessel steel had not been available for this analysis. Instead, a simple rule of thumb was used to estimate if the predicted stress peaks are tolerable. If the local stresses exceed the yield strength in operation, the material will experience plastic deformations there. Under cold, stress free conditions, this produces residual stresses with opposite sign up to the yield strength. As long as the local stress amplitude is less than twice the yield strength, the local stress-strain cycles remain purely elastic after a few initial cycles, thus cyclic plastic deformations are avoided. The steel is expected to stand a large number of such cycles even if the vessel is exposed to some neutron fluence.

The stress distribution in the upper part for steady-state condition before normal shut-down can be seen in Fig. 3.13. The critical region is at the inlet with maximum stresses of about 400 MPa. This point MAX was further evaluated during the transient analyses. In case of normal shut-down, the minimum and maximum temperature and the maximum stress history are shown in Fig. 3.14. The temperature is decreasing slowly as expected; at 18000 seconds, the minimum is close to that of the feedwater. The high stresses are due to the large temperature gradients at the inlet. Thermal shielding of the inlet nozzle by the backflow limiter, simulated by smaller heat transfer coefficients (HTC) in this region, did not reduce these stresses significantly. This latter case is called the modified HTC case in Fig. 3.14. But it shall be noted that, due to the smaller HTC, heat conduction is the dominating effect at the inlet in contrast to the preceding analysis with larger HTC where convection is dominating.

The results for the fast shut-down can be seen in Fig. 3.15. The critical region is the inlet opening again so that the point MAX is evaluated again. Within the first 720 seconds, the maximum stress increases to around 700 MPa. Afterwards, due to the fast pressure drop to 0.5 MPa and the decrease of the temperature gradient, the maximum stress decreases again.

Fig. 3.13: Steady state equivalent stresses in MPa of the RPV around the feedwater inlet flange at 20% load [13]

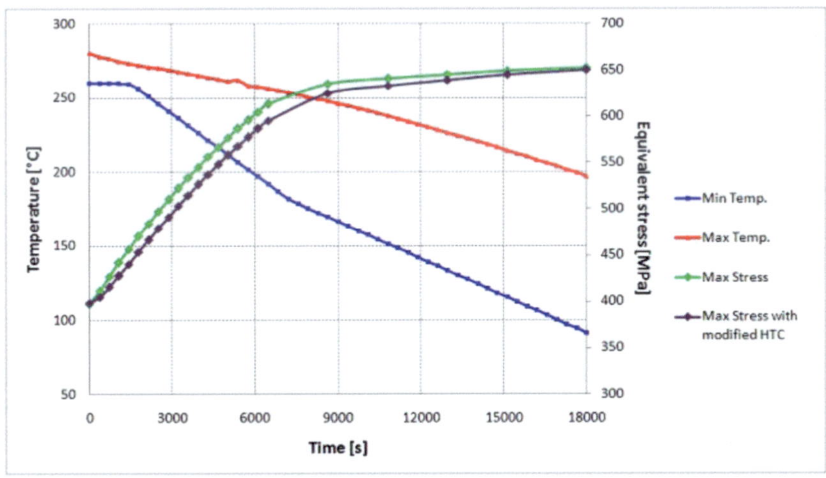

Fig. 3.14: Transient analysis of temperatures and equiv. stresses of the RPV around the feedwater inlet flange, assuming a normal shut-down from 20% load [13]

The steady state stress distribution near the steam outlet flanges is shown in Fig.3.16. Circles indicate regions of maximum stresses; these are: the support strut of the steam plenum (red circle in the centre), the inside of the outlet opening (orange circles) and the transition from the outlet flange to the vessel (yellow circles). In case of normal shut-down, the temperature and stress history are shown in Fig. 3.17. The maximum stress – located at the support strut of the steam plenum – is increasing and exceeds 1200 MPa. This is due to the sharp corner between the strut and the vessel and must be avoided by using appropriate radii here. Beside this local stress peak, the remaining stresses reach a maximum of 630 MPa which is plotted with a green line in Fig. 3.17. During fast shut-down, Fig. 3.18, the stress peak at the corner of the strut is repeated again. Assuming that this peak can be easily avoided by suitable radii, the maximum stress of the remaining body occurs at the yellow circles shown in Fig. 3.16 (it reaches almost 900 MPa).

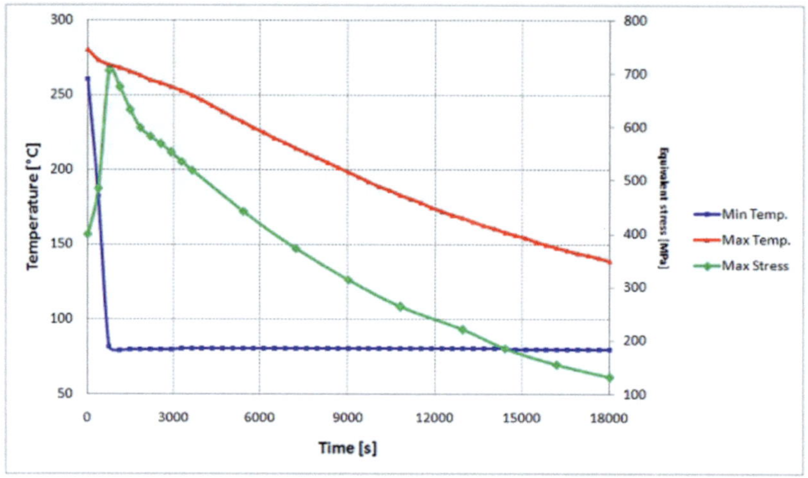

Fig. 3.15: Transient analysis of temperatures and equiv. stresses of the RPV around the feedwater inlet flange, assuming a fast shut-down from 100% load [13]

Fig. 3.16: Steady state equivalent stresses in MPa of the RPV around the steam outlet flange at 20% load [13]

110

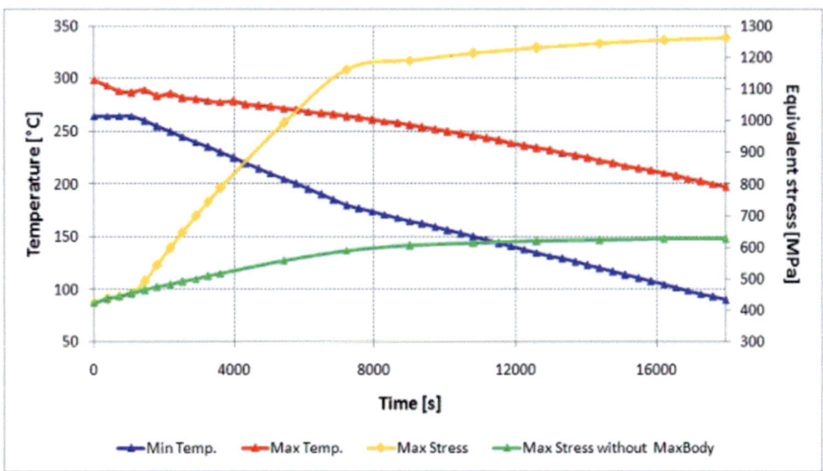

Fig. 3.17: Transient analysis of temperatures and equiv. stresses of the RPV around the steam outlet flange, assuming a normal shut-down from 20% load [13]

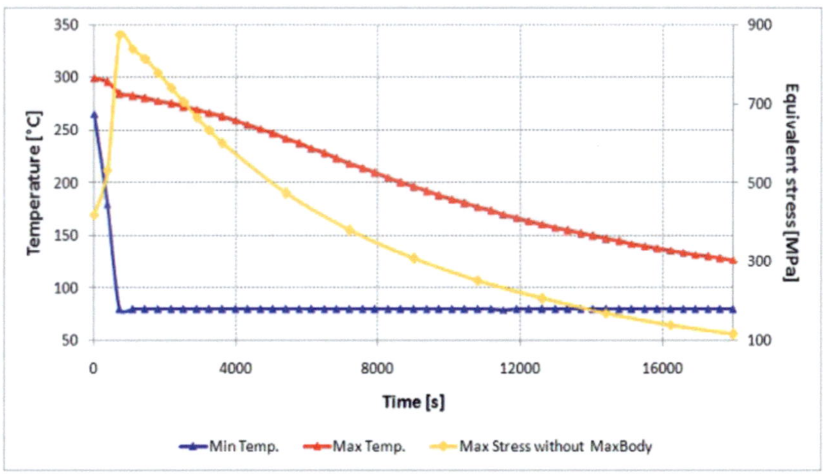

Fig. 3.18: Transient analysis of temperatures and equiv. stresses of the RPV around the steam outlet flange, assuming a fast shut-down from 100% load [13]

The outer steam flange was modelled separately. It is exposed to superheated steam and the thermal boundary conditions were estimated as follows (see Fig. 3.19 for explanation):

- Body 1 and 2: initial temperature 280°C, initial HTC 500 W/m^{2}K
- Body 4: initial temperature 280°C, initial HTC 3,000 W/m^{2}K
- Body 3 and 5: initial temperature 500°C, initial HTC 10,500 W/m^{2}K

Design details of the extractable steam pipe and the flow structure around it are shown in Fig. 3.20. Only bodies 5 and 3 are exposed to steam, whereas the other volumes are purged with feedwater.

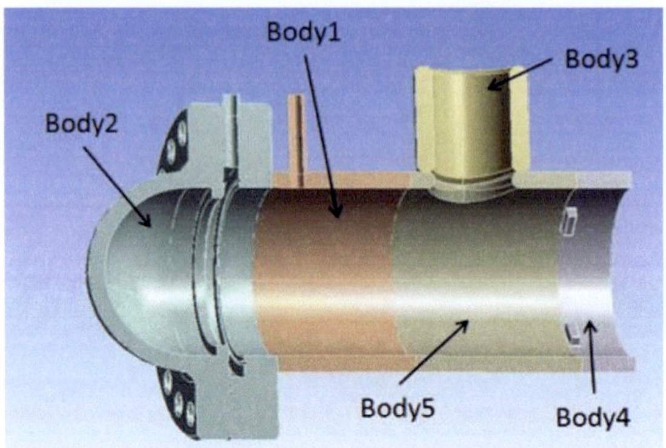

Fig. 3.19: Heat transfer areas of the thermal and stress analysis of the outer steam flange [13]

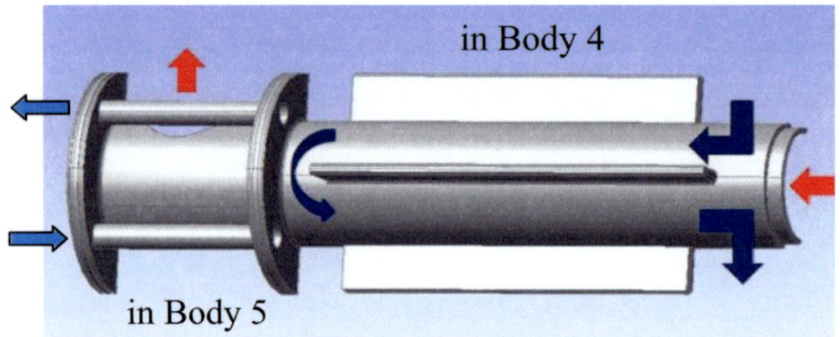

Fig. 3.20: Design details of the extractable steam pipe with fins

Two hypothetical transients were assumed exemplarily:

1. A case called "normal cool-down": at the initial steady-state condition, the pressure is 28.75 MPa, the mass flow rate is 100% of the full power operating conditions, the temperature of the fluid is 280°C if the part is exposed to feedwater and 500°C if it is exposed to superheated steam. During the transient phase, the pressure and heat transfer coefficients remain constant, while the temperature of the steam drops from 500°C to 400°C in 720 seconds and stays constant afterwards.

2. The case called "fast shut-down": the starting steady-state conditions are the same as before. During the transient, the pressure drops from 28.75 MPa to 0.5 MPa and the temperature from 500°C or 280°C, respectively, to 80°C in 720 seconds each. The heat transfer coefficients at the inside of the flange are reduced to 50% of the nominal value.

The material of the flange has been selected to be X 10 CrMoVNb 9 1 (P91).

The steady state temperatures in body 5 at the beginning of each transient are shown in Fig. 3.21. Temperature differences of around 100°C arise from thermal conduction to the adjacent pipe elements which are cooled down to less than 300°C. The maximum peak stress during steady state of 720 MPa occurs at the outlet nozzle, as indicated in Fig. 3.22 with a red circle. This is rather a consequence of the internal pressure which requires further design optimization to avoid fatigue at this point.

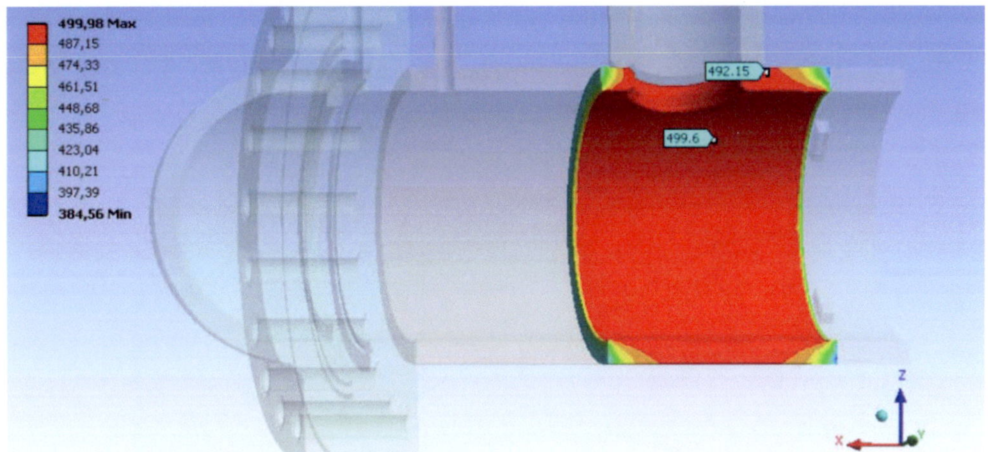

Fig. 3.21: Steady state temperature distribution in °C in body 5 of the outer steam flange [13]

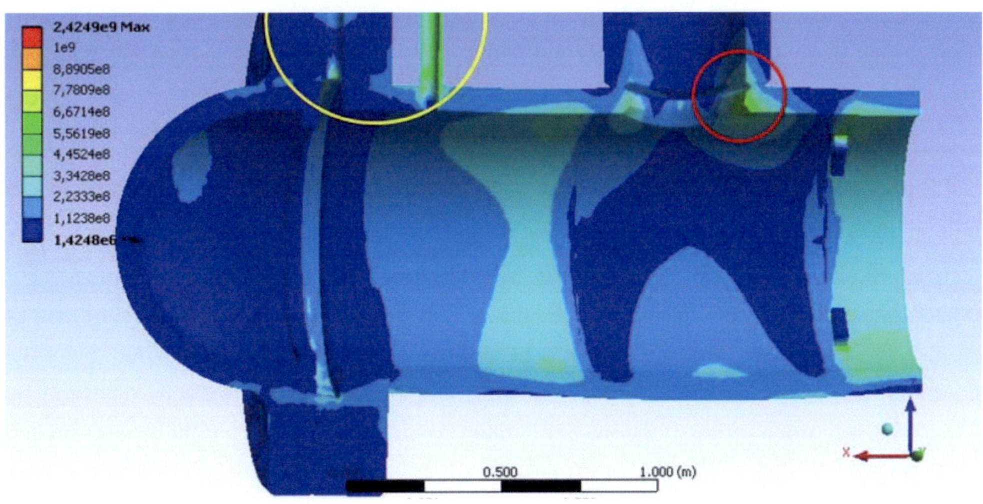

Fig. 3.22: Steady state equivalent stress distribution in Pa in the outer steam flange [13]

During the transients, the temperatures and stresses decrease rapidly, thanks to the thin walled structure of the outer steam flange. We see in Fig. 3.23 that the peak stress is hardly increasing, even during the fast shut down transient, as the temperature differences increase only to around 200°C at maximum after 720s, but the inside pressure decreases simultaneously.

The equivalent stress distribution after 540 s is shown in Fig. 3.24. At this time step, tensile stresses up to 600 MPa are distributed rather uniformly over the inner surface of body 5. (The peak stress indicated at the left supply line to the hydraulic piston is a consequence of an erroneous mechanical constraint there and should be ignored.)

The fatigue limit of all materials considered here is estimate as twice the yield strength, as explained above. Material data are given in Table 3.3.

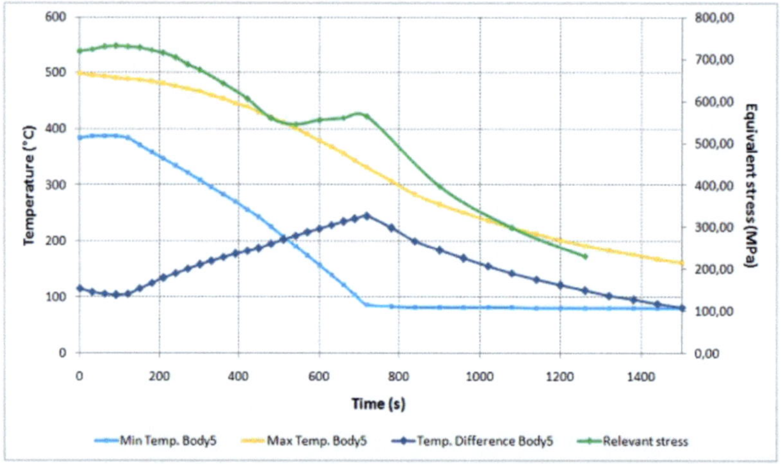

Fig. 3.23: Stress and temperature history in the outer steam flange during fast shut-down [13]

Comparing the predicted peak stresses with 2x the yield strength, as an estimate of the fatigue limit, we come to the following conclusions:

If sharp notches are carefully avoided by adding suitable radii, the thick walled RPV exceeds these limits by 3% in case of the normal shut-down transient, and by 43% in case of the fast shut-down one. The latter stress peak can be reduced if the transition from the outlet flange to the cylindrical vessel is designed with larger radius. It still exceeds this limit by 11%, however, at the inside of the inlet and outlet openings which can hardly be reduced any more (Figs. 3.13 and 3.16). Use of proper fatigue data and a minimization of temperature gradients during transients are required to avoid cracks in these locations.

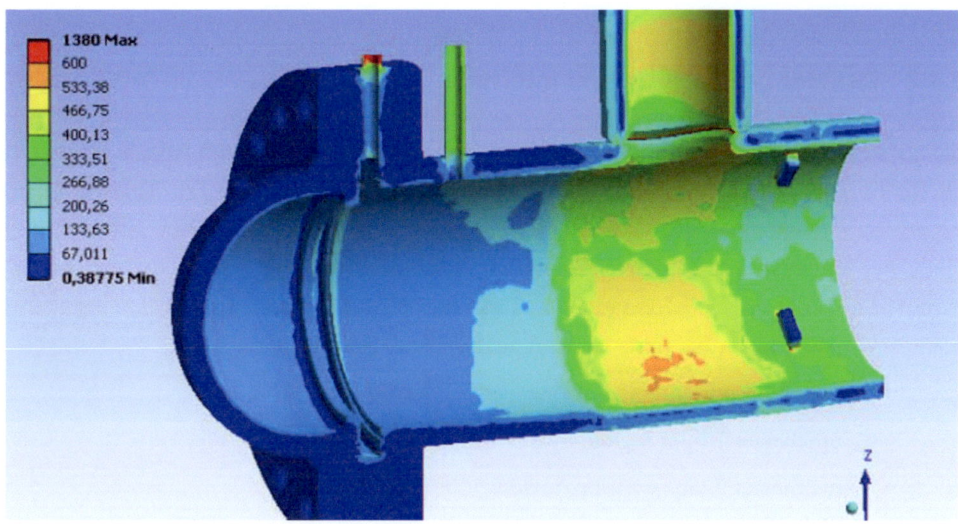

Fig. 3.24: Transient equivalent stresses in MPa in the outer steam flange, 540 s after the fast shut down [13]

Material	Temperature	$R_{p0.2}$	2x $R_{p0.2}$
20 MnMoNi 5 5	350°C	315 MPa	630 Mpa
X 10 CrMoVNb 9 1 (P91)	300°C	360 MPa	720 Mpa
X 10 CrMoVNb 9 1 (P91)	400°C	340 MPa	680 Mpa
X 10 CrMoVNb 9 1 (P91)	500°C	300 MPa	600 MPa

Tab. 3.3: Yield strength of the assumed vessel materials [13]

The outer steam flange experiences similar temperature differences even though it is hotter by about 200°C. The maximum peak stress of 775 MPa during normal cool-down transient exceeds the estimated limit by around 30%. However, thanks to the thin walled structure, these stresses are not higher in case of the assumed fast shut-down transient. The peak stresses are rather a consequence of the internal pressure and can be minimized by a smoother transition from the outlet flange to the steam line.

As a general conclusion, the predicted peak stress levels are high but still in a reasonable range so that an acceptable lifetime can be expected after careful design optimization.

3.4.2 Reactor pressure vessel internals

Steady state temperatures, stresses and deformations of the steam plenum have been analyzed by Redon [14], which have been summarized also by Fischer et al. [9]. This component experiences temperature differences ranging from 280°C on its contact surface to the feedwater temperature up to 500°C at the steam outlet of the second superheater, which is expected to cause larger deformations and stresses. Redon estimated the following thermal boundary conditions for his finite element analysis with ANSYS:

- Evaporator outlet region, inside: 390°C, 13900 W/m^{2}K,

- Superheater 1 inlet region, inside: 390°C, 12600 W/m^{2}K,

- Superheater 2 outlet region, inside: 500°C, 3200 W/m^{2}K,

- Upper surface, outside: 280° C, 500 W/m^{2}K,

- Lower surface, outside: 302°C, 500 W/m^{2}K.

The resulting steady state temperature distribution is shown in Fig. 3.25. The temperature differences of this component are almost as large as those of the surrounding fluids.

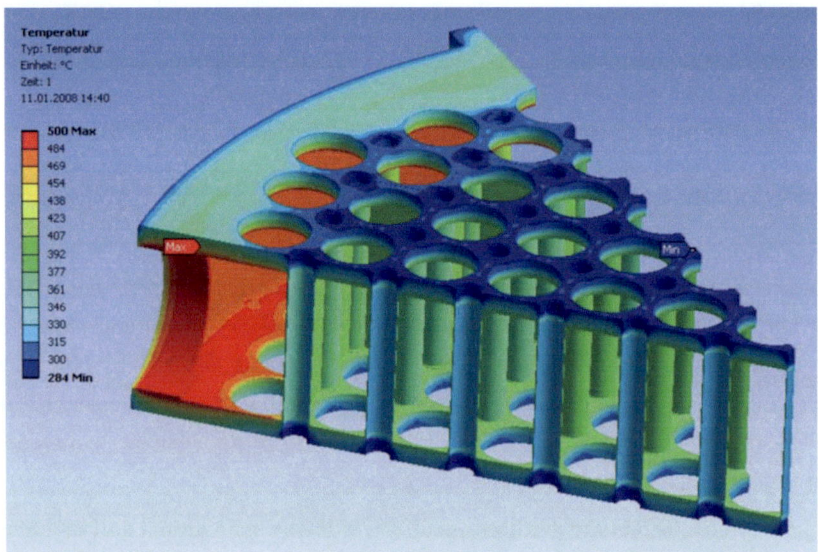

Fig. 3.25: Steady state temperature distribution in °C of the steam plenum [9]

The thermal deformations of the steam plenum under these thermal loads are shown in Fig. 2.49, where their consequences on the leakage of feedwater into the steam plenum have already been discussed. The maximum steady state stresses and the allowable stress level of

SS 316 L at these temperatures are listed in Tab. 3.4. The stresses predicted by the finite element analysis have been categorized according to the KTA Guidelines [15]. The stresses are acceptable, but the improved material 1.4970 would leave more margins for transients.

Location	Stress category	Max. stress [MPa]	Allowable stress SS 316 L [MPa]
Steam outlet nozzle	Membrane (I)	148	198
Steam outlet nozzle	Membrane+bending (II)	224	297
Steam outlet nozzle	Membrane+bending+thermal (III)	585	594
Assembly penetrations	Peak (IV)	601	662

Tab. 3.4: Maximum stress level of the upper plenum under steady state conditions [9]

Stresses and deformations of the core support plate with its lower mixing plenum have been analyzed by Koehly et al. [16] for steady state conditions. In particular, the outlets of moderator water from the reflector to the lower mixing plenum had been a concern since they are perforating the core support plate at its outer periphery. Maximum peak stresses were up to 620 MPa, as shown in Fig. 3.26, which are still acceptable according to Tab. 3.4.

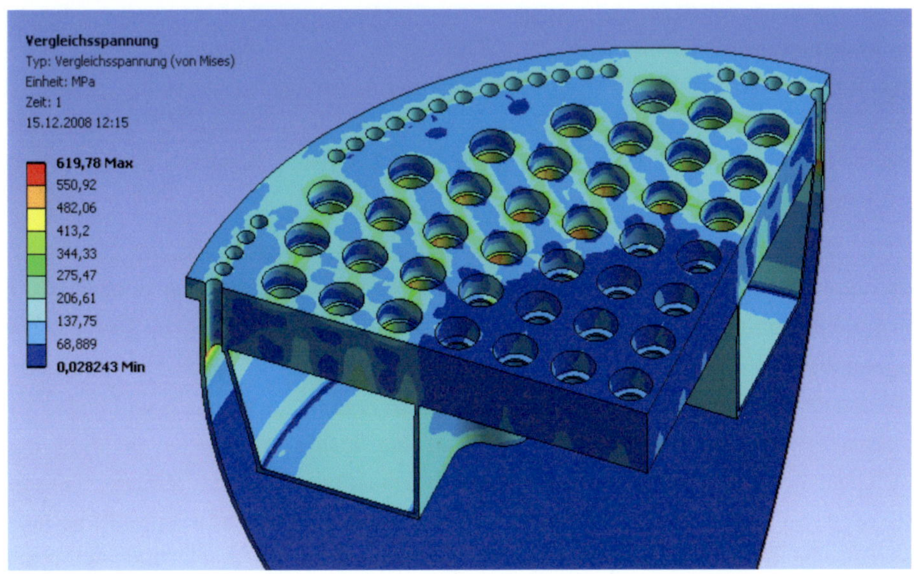

Fig. 3.26: Equivalent stresses in MPa of core support plate and lower mixing plenum under steady state conditions, in MPa [16]

3.5   In-core instrumentation

An in- core instrumentation for the HPLWR has been designed by Koehly et al. [17]. It consists of three different measurement systems, of which two measurement systems are foreseen for the power distribution in the core: The continuously working neutron-flux density distribution and the discontinuously working aeroball measurement system. The third measurement system consists of thermo-couples for measuring the different coolant temperatures within the core. Because of the new geometry and design of the HPLWR in conjunction with the different design of the fuel assembly clusters and control rods, the existing PWR technology had to be adapted to the new conditions.

For measuring and controlling the power distribution in the core, the neutron-flux density distribution measurement system has been proposed. The main task of this system is to monitor the nuclear power distribution. With this system, the power density within the core can be indicated and limited. The system works continuously during the reactor operation by means of 12 Power Distribution Detector (PDD) fingers distributed over the core in 180° symmetry arrangement. Each detector finger has 6 axially distributed self-powered neutron detectors. They consist of a Cobalt-59 emitter, which generates an electric current through nuclear reactions. These prompt signals are directly proportional to the thermal neutron flux. The measured power density range for PWRs lies between 7 W/cm to 590 W/cm and works therein with detector linearity from 10 % nominal power up to reactor trip by High Level Power Density (HLPD) protection. These values should be applicable for the HPLWR. The detector emitters have to be exchanged about every 2-3 years. Therefore, the detector fingers have to be replaceable.

The second measurement system for neutron-flux density is the aeroball probe measurement system. This system is mainly used for measuring the three dimensional power density distribution and calibration of the neutron detectors. Each aeroball finger consists of an outer instrumentation finger and two interleaved tubes, one for guiding the Vanadium steel balls and the other one for the carrier gas. For measurement, the aeroballs will be blown from a measurement room outside of the reactor but inside the reactor building into the instrumentation finger within the core. After an irradiation of about 3 minutes, the aeroballs will be blown back to the measurement table, where the activity of a ball will be measured and decoded to the 3-dimensional power distribution of the core. The total measuring procedure and flux mapping will need 10 minutes. The aeroball measurement system supplies the data for monitoring, calibration and limit value installation of the PDD system.

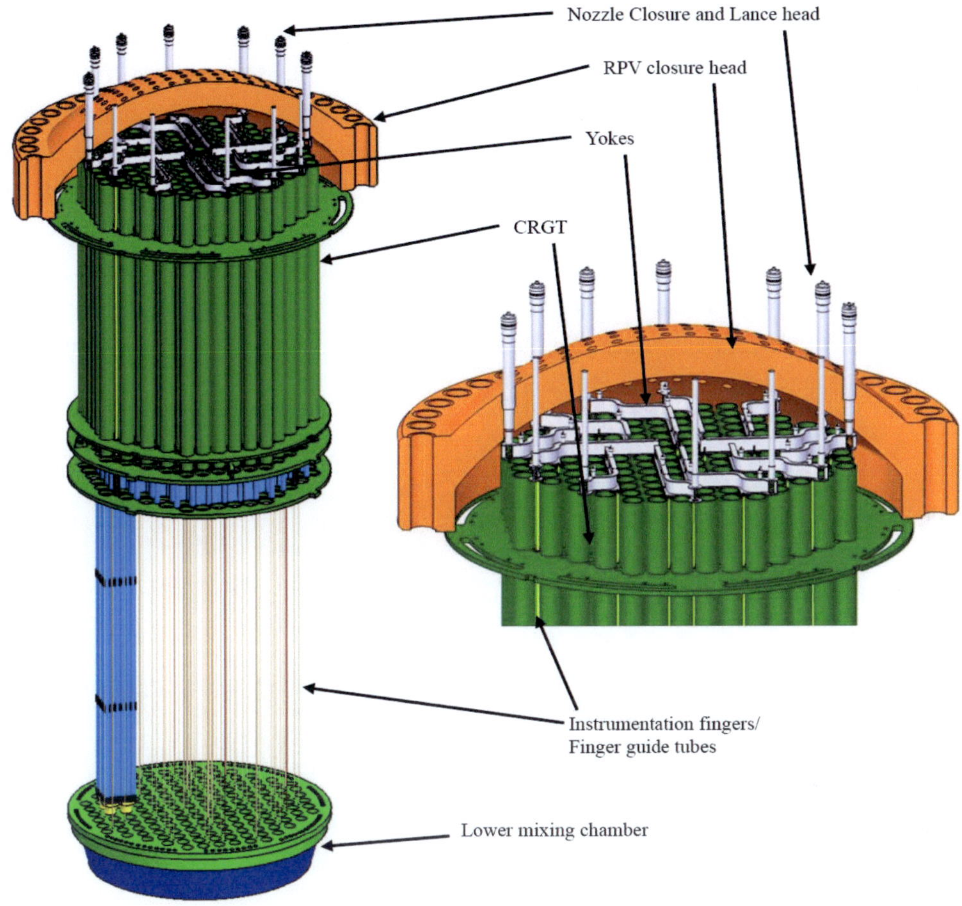

Fig. 3.27: Integration of the HPLWR in-core instrumentation [17]

The third system of the in-core instrumentation, which has been integrated into the HPLWR concept, is a temperature measurement system at core outlet, which is here, in particular, the outlet temperatures of each heat-up step of the HPLWR, namely Evaporater, Superheater 1 and Superheater 2, measured by thermo-couples. There are 12 temperature measuring fingers, 4 for each of the three heat-up steps. Because of the different positions of outlet areas of the 3 heating steps, a different length of the measurement fingers is required. For measuring the outlet temperatures of Evaporator and Superheater 2, the measurement should be foressen at the axial height of the windows in the fuel assembly head pieces. The thermo-couples are installed within the short instrumentation fingers and have an indirect connection through the guide tubes to the steam outside. The measurement of Superheater 1 outlet temperature has to be placed within the lower mixing plenum. Therefore, long instrumentation fingers are needed.

Fig. 3.27 shows the integration of the in-core instrumentation into the HPLWR and, in particular, the lead through in the closure head. The yokes of the instrumentation lances lie on the top of the control rod guide tubes (CRGT). Instrumentation fingers are fixed at the yokes and extended downwards either to the core support plate or to the steam plenum (depending on the finger length). The shafts will be guided at an outer radius of the RPV, upwards through the RPV closure head, by a leak- and pressure-tight weld and bolt connection. For refueling, the instrumentation lances have to be disconnected and closed with the main plug. After disassembling the RPV closure head, the instrumentation lances have to be withdrawn before disassembling the control rod guide tubes. Fig. 3.28 shows the arrangement of the core instrumentation lances from top view. There are 6 different lances arranged in 180° symmetry within the core.

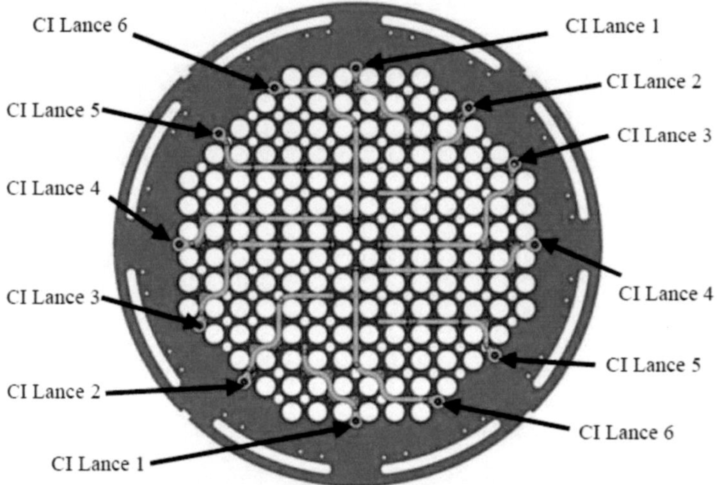

Fig. 3.28: Arrangement of core instrumentation (CI) lances; view from top [17]

According to the core power distribution described in Chapter 2, the core instrumentation could be positioned as shown in Fig. 3.29. To avoid a difficult and complicated system in sealing and guidance, the detector fingers have been placed between the corners of four fuel assembly clusters. At these positions, only few additional penetrations and guidance will be needed and existing penetrations in the steam plenum can be used. Overall, there are 12 instrumentation lances. On the outermost radius, there are the 12 shafts which guide all cables and tubes through the closure head to the outside of the reactor. A yoke is attached to each shaft. The different instrumentation fingers are fixed at the yokes and guided downwards into the core to the measurement positions. Four kinds of instrumentation fingers have been proposed: Two of them are the neutron detector and the aeroball measuring tubes, reaching down below the active core length. The other two lances are the temperature measurement lances, which are needed in two different sizes: a long lance for the temperature measurement in the lower mixing chamber for the outlet of superheater 1 and a short lance for temperature measurement within the steam plenum at the outlet of the evaporator and superheater 2. Thus, there are 6 different assembled instrumentation lances distributed with 180° symmetry in the core. Each of them has different instrumentation fingers fixed at the yoke. Fig. 3.29 shows the proposed arrangement in the HPLWR core. The different heat up steps are indicated with different colors. A blue line indicates the inner wall of the lower mixing chamber. In total, there are 12 temperature and 36 aeroball measuring fingers, as well as 12 neutron detector fingers.

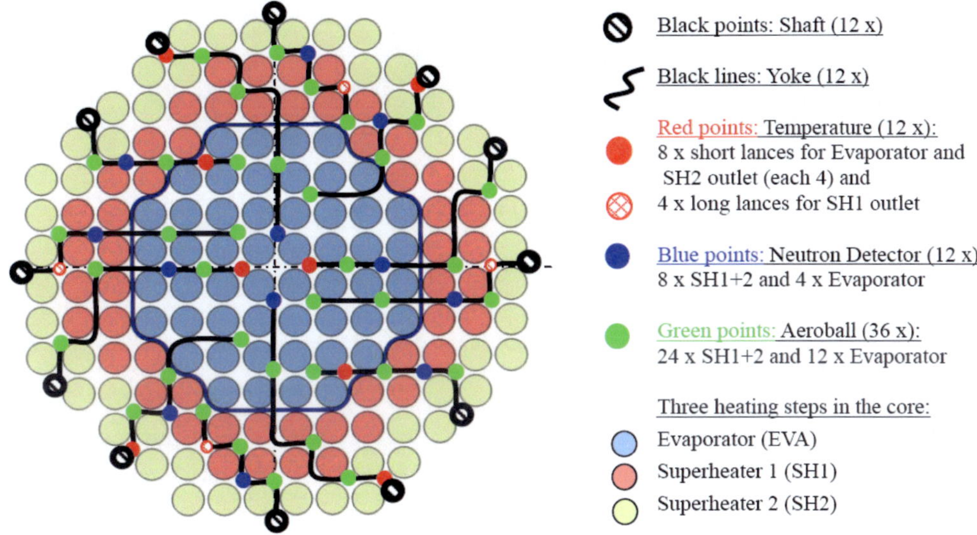

Fig. 3.29: Position of instrumentation fingers in the HPLWR core [17]

Some design details of the penetrations through the steam plenum and of measurement positions inside the steam plenum are shown in Fig. 3.30. The stiffening tubes, connecting upper and lower plate of the steam plenum, are used to guide the long measurement fingers. A sealing is needed at the upper end of these tubes, while the lower end remains open. In case that a measurement needs to be taken inside the steam plenum, these stiffening tubes are simply opened to the steam and a sealing is needed on both ends.

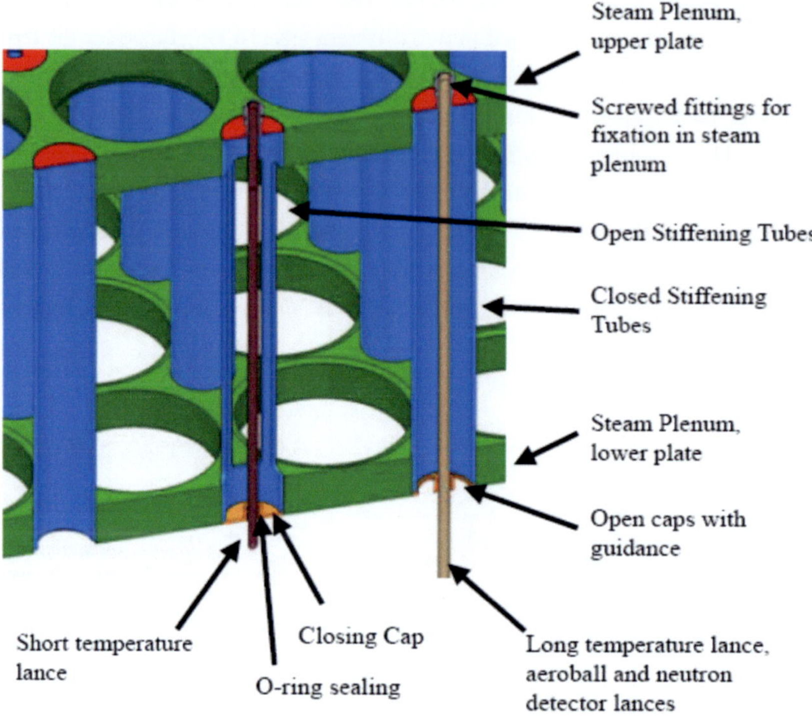

Fig. 3.30: Penetration of temperature lances through the steam plenum and example of a temperature measurement inside the steam plenum [17]

As the lances to the lower mixing plenum are running between the corners of the fuel assembly clusters, the outer spacer pads of each cluster, Fig. 2.48, need to be modified such that they leave enough space for the lances. Fig. 3.31 shows an example.

Fig. 3.31 Core instrumentation lances between the corners of the assembly clusters (left) and modification of the spacer pads (right) [17]

3.6 Fluence analysis

The lifetime of a nuclear power plant is ultimately restricted by the degradation of the mechanical properties of the reactor pressure vessel, which is mainly caused by the fast neutrons. According to the experiences, this degradation is determined by the time integral of the displacements per atom (dpa) rate. For the proper analysis of this quantity, the detailed operational history needs to be known, so that a large number of shielding calculations can be performed. This is mostly done by specifying the neutron source in the pins of the outer fuel assemblies and treating the transport of neutrons to the pressure vessel by a Monte Carlo code. To save the required computer time, a new procedure, specifying the neutron source on the outer surface of the reactor core was developed formerly. This procedure was applied by Keresztúri et al. [18] to the VVER-440 units of Paks NPP, Hungary. Its accuracy was tested by comparison with neutron dosimetric measurements at these units. The same procedure was applied to the HPLWR reactor. The source to be evaluated on the outer surface of the core and the neutron transport outside the core were calculated by the Monte Carlo code MCNP.

The analyses for the HPLWR were performed by Hordósy et al. [19] in a 45 degree symmetry sector. The source was specified on the outward surfaces of the 19 outermost assemblies. Because the analyses of the core power distribution with KARATE, Chapter 2.4, were using 39 axial nodes, these surfaces were divided axially into 39 "pages" with equal height again. Different energy spectra, calculated by KARATE, were given on each page. The horizontal cross section of the model is shown in Fig.3.32. Two different design options for the reflector were studied: a solid steel reflector and a reflector design with water channels as shown in Figs. 2.8 and 3.32.

Both configuration calculations were performed for the specified lifetime of 60 years. The main characteristics were as follows:

Configuration 1:

- First cycle
- 325 day cycle length
- Downward gap flow between the assembly clusters
- Steel reflector

Configuration 2:

- Equilibrium cycle

- 355 day cycle length

- Upward gap flow between the assembly clusters

- Steel-water reflector

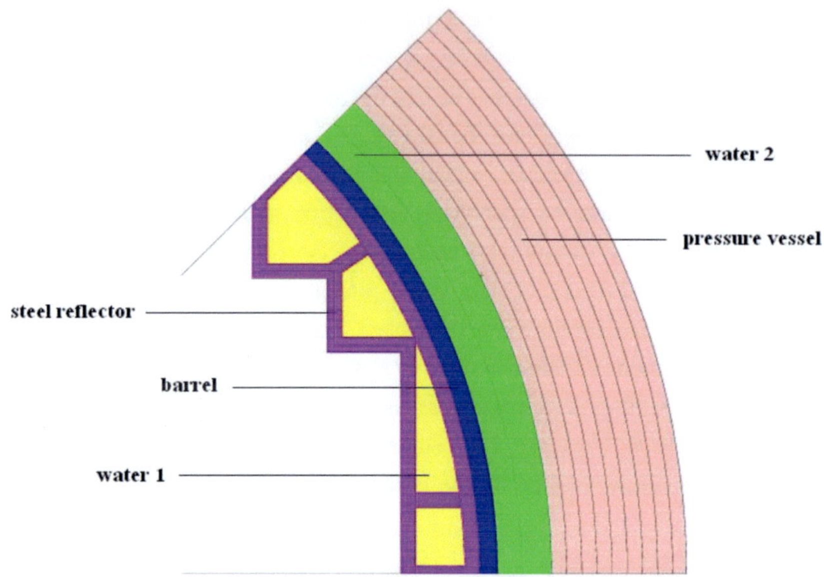

Fig. 3.32: Horizontal cross section of the model for fluence analysis [19]

The source was evaluated approximately for 2 weeks time steps for a cycle. Monte Carlo calculations were performed to determine the dpa distribution in the pressure vessel for these steps and the results were integrated.

Based on the results of the calculations, the following conclusions can be drawn:

The radial power distribution of the three pass core concept is advantageous to RPV fluence.

Introducing water channels in the reflector steel affects only slightly the maximum fluence of the RPV.

The calculated HPLWR RPV damage is close to current PWR damage and significantly lower than VVER-440 damage with 50 year lifetime possibility.

3.7 Discussion

Even though pressure and temperatures of the HPLWR are higher than in a pressurized water reactor, many of the proven PWR design concepts could be applied again for this new concept. The arrangement of control drives and control rod guide tubes is even almost identical to the PWR. Major differences, however, are caused by the meandering flowpath of moderator water and of the coolant, which does not only change the core design, but also the design of several components around the core. It has not been easy in the beginning to obtain a stable flow configuration for such a flowpath, but sufficiently high flow velocities, which overcome the problem of flow reversal by buoyancy effects, could finally be achieved with orifices and flow restrictions. These, in turn, were increasing the total pressure drop of the coolant on its way from inlet to the outlet of the reactor. However, as recirculation pumps are not foreseen for the HPLWR, even a total pressure drop of 1 MPa would be of minor importance for the overall plant efficiency and for plant erection costs. The concept shown here stays below this limit.

A great challenge has been to design the internals of the pressure vessel such that they can freely expand under the increased temperature differences, but to seal each component against the others such that cold feedwater cannot penetrate into the hot steam. It is still an open question, how close these sealing systems can be built, and how durable they will be stay under long term operation.

Detailed analyses of primary and thermal stresses of the RPV and of larger components inside confirmed that the stress level is expected to be tolerable, thanks to the thicker pressure boundary, a consequent thermal shielding of thick walled structures from superheated steam, and from minimization of mechanical constraints. Most of the predicted peak stresses should be avoidable with some further design optimization of radii in notches.

The core design with its low power assemblies of the 2nd superheater, next to the core periphery, is advantageous for the lifetime of the pressure vessel. These assemblies are shielding the high neutron flux of the central assemblies and minimize the displacements in the vessel material.

The backflow limiter at the feedwater inlet is certainly minimizing the loss of coolant in case of a break of a feedwater line. The next chapter will show, however, that the coolant inventory in the pressure vessel is not an adequate criterion for coolability, but a sufficient coolant mass flow through the core must be maintained instead. The effect of the backflow limiter will need to be discussed again in context with transient analyses.

3.8 References

[1] K. Kataoka, S. Shiga, K. Moriya, Y. Oka, S. Yoshida, H. Takahashi, Progress of Development Project of Supercritical-water Cooled Power Reactor, Proc. ICAPP '03, Paper 3258, Cordoba, Spain, May 4-7, 2003

[2] J. Buongiorno, The Supercritical Water Cooled Reactor: Ongoing Research and Development in the U.S., Proc. ICAPP '04, Paper 4229, Pittsburgh, USA, June 13-17, 2004

[3] J. Buongiorno, P. MacDonald, Supercritical Water Reactor (SCWR), Progress Report for the FY-03 Generation-IV R & D Activities for the Development of the SCWR in the U.S., INEEL/EXT-03-01210, Idaho National Engineering and Environmental Laboratory (INEEL), September 2003

[4] D. Bittermann, D. Squarer, T. Schulenberg, Y. Oka, P. Dumaz, R. Kyri-Rajamäki, N. Aksan, Cs. Maráczy, A. Souyri, Potential Plant Characteristics of a High Performance Light Water Reactor (HPLWR), Proc. ICAPP '03, Cordoba, Spain, May 4-7, 2003

[5] K. Fischer, J. Starflinger, T. Schulenberg, Conceptual design of a reactor pressure vessel and its internals for a HPLWR, Proc. ICAPP '06, Paper 6098, Reno, NV, USA, June 4-8, 2006.

[6] C. Koehly, T. Schulenberg, J. Starflinger, Design concept of the HPLWR moderator flow path, Proc. ICAPP '09, Paper 9187, Tokyo, Japan, May 10-14, 2009

[7] K. Fischer, T. Schneider, T. Redon, T. Schulenberg, J. Starflinger, Mechanical design of core components for a High Performance Light Water Reactor with a three pass core, Proc. GLOBAL 07, Boise, Idaho, USA, Sept. 9-13, 2007

[8] J. Hofmeister, T. Schulenberg, J. Starflinger, Optimization of a Fuel Assembly for a HPLWR, Proc. ICAPP '05, Paper 5077, Seoul, Korea, May 15-19, 2005

[9] K. Fischer, T. Redon, G. Millet, C. Koehly, T. Schulenberg, Thermo-mechanical stress and deformation analysis of a HPLWR pressure vessel and steam plenum, Proc. ICAPP '08, Paper 8050, Anaheim, CA, USA, June 8-12, 2008

[10] K. Fischer, E. Laurien, A.G. Class, T. Schulenberg, Design and optimization of a backflow limiter for the High Performance Light Water Reactor, Proc. GLOBAL 07, Boise, Idaho, USA, Sept. 9-13, 2007

[11] C. Kunik, B. Vogt, T. Schulenberg, Flow phenomena in the gap volume between assembly boxes, 4th International Symposium on Supercritical Water-Cooled Reactors, Paper No.04, Heidelberg, Germany, March 8-11, 2009

[12] H. Foulon, A. Wank, T. Schulenberg, CFD analysis of feedwater flow in the HPLWR pressure vessel, Int. Students Workshop on HPLWR, Karlsruhe, Germany, March 31 to April 3, 2008

[13] T. Reiss, H. Foulon, A. Wank, T. Schulenberg, Transient stress analysis of the HPLWR pressure vessel, 4th International Symposium on Supercritical Water-Cooled Reactors, Paper No. 39, Heidelberg, Germany, March 8-11, 2009

[14] T. Redon, Auslegung der oberen Mischkammer eines Reaktors mit überkritischen Dampfzuständen, Diplomarbeit, Institute for Nuclear and Energy Technologies, Forschungszentrum Karlsruhe, May 2007

[15] KTA-Geschaeftsstelle c/o Bundesamt fuer Strahlenschutz (BfS), "Reactor Pressure Vessel Internals," Part No. 3204, Kerntechnischer Ausschuss (KTA), Germany, 1998

[16] C. Koehly, T. Schulenberg, J. Starflinger, HPLWR reactor design concept, 4th International Symposium on Supercritical Water-Cooled Reactors, Paper No.37, Heidelberg, Germany, March 8-11, 2009

[17] C. Koehly, W. Meier, J. Starflinger, Integration of in-core instrumentation into HPLWR, Proc. ICAPP 10, Paper 10167, San Diego, CA, USA, June 13-17, 2010

[18] A. Keresztúri, Cs. Hegedűs, Gy. Hegyi, G. Hordósy, M. Makai, M. Telbisz, Karate - A code for VVER-440 Calculation, Proc. of the 5th Symposium of AER, p.403-421, 1995

[19] G. Hordósy, Gy. Hegyi, A. Keresztúri, Cs. Maráczy, Fluence analysis of the reactor pressure vessel wall, unpublished, 2009

4 Containment and Safety Systems

Since the HPLWR is considered to be a long term development project which is not expected to be realized in near future, it is somewhat difficult to foresee the requirements which will be appropriate at that time. As a general guideline, the requirements known from the Generation IV initiative [1], and more specific the European Utility Requirements (EUR) [2], which are currently considered to be most advanced and most complete in Europe, were taken into account for HPLWR design. The latter requirements are more useful for practical design purposes.

With respect to reactor safety, there are 3 major Generation IV goals to be considered, which are:

(1) Generation IV nuclear energy systems operations will excel in safety and reliability.

(2) Generation IV nuclear energy systems will have a very low likelihood and degree of reactor core damage.

(3) Generation IV nuclear energy systems will eliminate the need for offsite emergency response.

These general goals were defined in the Generation IV program [1] in more detail, and the essential parts of it is referenced in the following sections.

Safety goal (1) aims at increasing operational safety by reducing: the number of events, equipment problems, human performance issues that can initiate accidents or cause them into more severe accident. It also aims at achieving increased nuclear energy systems reliability that will benefit their economics. Appropriate requirements and robust designs are needed to advance such operational objectives and to support the demonstration of safety that enhance public confidence. During the last two decades, operating nuclear power plants have improved their safety levels significantly. At the same time, design requirements have been developed to simplify their design, enhance their defense-in-depth in nuclear safety, and improve their constructability, operability, maintainability, and economics. Increased emphasis is being put on preventing abnormal events and on improving human performance by using advanced instrumentation and digital systems. Also, the demonstration of safety is being strengthened through prototype demonstration that is supported by validated analysis tools and testing, or by showing that the design relies on proven technology supported by ample analysis, testing, and research results. Radiation protection is being maintained over the total system lifetime by operating within the applicable standards and regulations. The concept of keeping radiation exposure as low as reasonably achievable (ALARA) is being successfully employed to lower radiation exposure. Generation IV nuclear energy systems must continue to promote the highest levels of safety and reliability by adopting established principles and best practices developed by the industry and regulators to enhance public confidence, and by employing future technological advances. The continued and judicious pursuit of excellence in safety and reliability is important to improving economics.

Safety goal (2) is vital to achieve investment protection for the owner/operators and to preserve the plant's ability to return to power. There has been a strong trend over the years to reduce the possibility of reactor core damage. Probabilistic Risk Assessment (PRA) identifies and helps prevent accident sequences that could result in core damage and off-site radiation releases and reduces the uncertainties associated with them. For example, the US ALWR Utility Requirements Document [3] requires the plant designer to demonstrate a core damage frequency of less than 10^{-5} per reactor year by PRA. This is a factor of about 10 lower in frequency by comparison to the previous generation of LWR energy systems. Additional means, such as passive features to provide cooling of the fuel and reducing the need for uninterrupted electrical power, have been valuable factors in establishing this trend. The evaluation of passive safety should be continued and passive safety features incorporated into Generation IV nuclear energy systems whenever appropriate.

The intent of safety goal (3) is, through design and application of advanced technology, to eliminate the need for off-site emergency response. Although its demonstration may eventually prove to be unachievable, this goal is intended to simulate innovation, leading to the development of designs that could meet it. The strategy is to identify severe accidents that lead to offsite radioactive releases, and to evaluate the effectiveness and impact on economics of design features that eliminate the need for offsite emergency response. The need for offsite emergency response has been interpreted as a safety weakness by the public and specifically by people living near nuclear facilities. Hence, for Gen IV systems a design effort focused on elimination of the need for offsite emergency response is warranted. This effort is in addition to actions, which will be taken to reduce the likelihood and degree of core damage required by the previous goal.

A comparison between the GENIV goals and the EUR was performed by Bittermann et al. [4], which showed, in general, that the Generation IV requirements are compatible with the top tier EUR document. This is an important observation, since by using the EUR as a guide for the detailed design of the HPLWR, it will also insure the conformity of the HPLWR with Generation IV goals.

Defense in depth is one of the important principles in all safety concepts of current reactors and it shall consequently be applied also for the HPLWR. Accordingly, save operation of the power plant shall be ensured by the following measures:

Normal operation (DBC1) controlled by operating systems

- Conservative design, reliability, availability
- Proven technology, quality assurance

Operational occurrences (DBC2, $>10^{-2}$/year) controlled by control and limitation features

- Surveillance, diagnostics
- Inherent safety, nuclear stability

Design basis accidents (DBC3/4, $>10^{-5}$/year) controlled by safety systems

- Redundancy, train separation
- Protection against internal and external hazards
- Qualification against accident conditions
- Automation (<30 min)
- Autarchy

Reactor state or event	Limited parameter	Limit	Reason for limit
Normal operation DBC1	Maximum cladding temperature	630°C	Cladding creep or corrosion
	Maximum linear heat rate	39 kW/m	Fission gas release, pressure inside fuel rod
Operational occurrence, DBC2	Maximum cladding temperature	850°C	Cladding buckling collapse
	Max. fuel centerline temperature	2800°C	Fuel melting
Design basis accidents, DBC3/4	Maximum cladding temperature	1200°C	Cladding embrittlement due to oxidation
	Maximum radially averaged pellet enthalpy	963 kJ/kg	Fuel fragmentation and dispersion

Table 4.1: Preliminary target data for safety system assessment

Design Extension 1 (DEC1): Multiple failure scenarios (e.g. station blackout, total loss of feedwater, loss of coolant accidents), severe external events (e.g. military or large commercial airplane crash)

- Diversified systems

- Design against external event loads

Design Extension 2 (DEC2): Severe Accidents

- Mitigative features

- Prevention of energetic consequences which could lead to large early containment failure (e.g. steam explosion, direct containment heating, global hydrogen detonation)

Some transient system analyses have been performed exemplarily for the HPLWR to show if the reactor design will enable such a defense in depth strategy, as will be described in chapter 4.3. As long as the cladding material and its strength and corrosion resistance have not yet been defined, acceptance criteria for the maximum fuel and cladding temperature can hardly be quantified. Instead, Table 4.1 shall provide some preliminary target data to assess the safety system design.

4.1 General strategy of the safety system

As the HPLWR has a once through steam cycle, in which steam from the core outlet is directly supplied to the high pressure turbines, it has many similarities with boiling water reactors (BWR). In general, therefore, similar safety system could be applied again. On a closer look, however, there is a basic difference in the coolant flowpath inside the reactor, which causes a different safety strategy, as discussed by Schulenberg and Visser [5]. To explain this, the simplified control systems of a BWR and the HPLWR are compared in Fig. 4.1. In a BWR, the feedwater pump is controlling the liquid level in the reactor pressure vessel, the steam pressure is controlled by the turbine governor valve, and the core power is either controlled by the control rods or by the speed of the recirculation pumps, indicated with blue circles in Fig. 4.1, left. The HPLWR, on the right hand side, does not include any recirculation loop. The feedwater pump can either control the steam temperature at the core outlet if the core power is controlled by the control rods, or it can control the core power if the steam outlet temperature is controlled by the control rods. Again, the steam pressure is controlled by the turbine governor valve in both cases.

This once through system is causing also a basic difference to the general strategy of the safety system. Having a closed coolant loop inside the reactor, the BWR can remove residual heat by natural convection, driven by the rising steam in the core and above, and the safety system has to ensure sufficient coolant inventory in the pressure vessel to keep the core covered with water. The HPLWR, on the other hand, can remove the residual heat only by forced convection inside the reactor, which may be driven by a natural convection loop outside, but the requirement for the safety system, in general, is to ensure sufficient coolant mass flow rate instead. Besides this difference, there are several common safety system requirements, which can be taken directly from BWR concepts without significant modifications. These are:

- The reactor shut down system by control rods or by a boron injection system as a second, independent shut down system.

- Containment isolation by active and passive containment isolation valves in each line penetrating the containment to close the third barrier in case of an accident.

- Steam pressure limitation by pressure relief valves.

- Automatic depressurization of the steam lines into a pool inside the containment through spargers to close the coolant loop inside the containment in case of containment isolation.

- A coolant injection system to refill coolant into the pressure vessel after intended or accidental coolant release into the containment.

- A pressure suppression pool to limit the pressure inside the containment in case of steam release inside the containment.

- A residual heat removal system for long term cooling of the containment.

The following sections will discuss the safety strategy using some examples of safety concepts which had been worked out recently for supercritical water cooled reactors.

A minimum set of safety systems which fulfill the above mentioned requirements is sketched in Fig. 4.2. The reactor shut down system is provided by shut down rods which can fall into the reactor from the top like in a pressurized water reactor (PWR), since separators and dryers are not complicating the design like in a BWR. The HPLWR design, Fig. 3.2, shows that control rod drives outside the reactor as well as control rod guide tubes inside can be taken from PWR design without significant modifications. In addition, a vessel with boron acid must be provided inside or outside the containment for redundant shut down under accidental conditions. Different from PWR control, however, this boron acid cannot be used to compensate excess reactivity during normal operation.

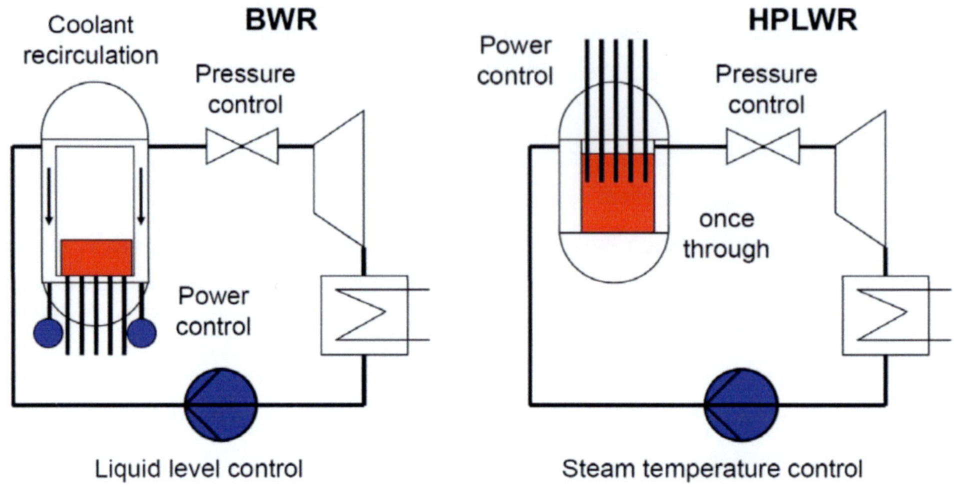

Fig. 4.1: Comparison of BWR and HPLWR general systems [5]

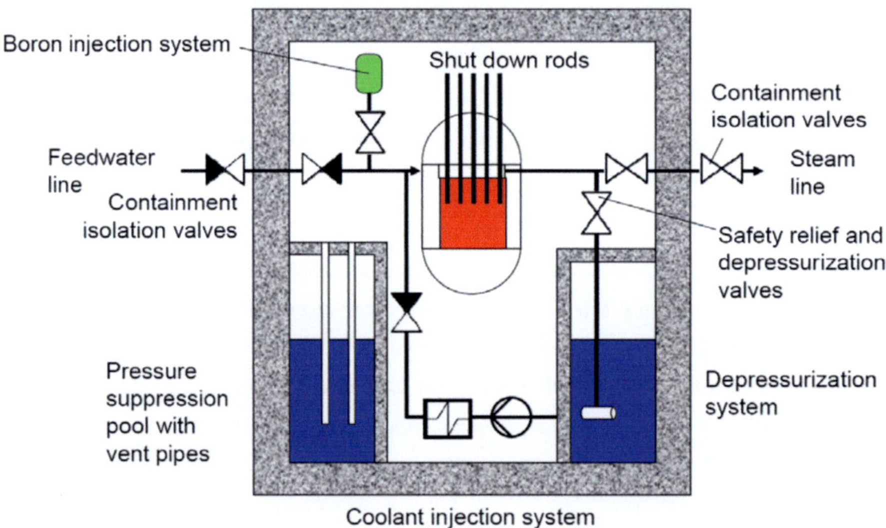

Fig. 4.2: Minimum set of safety systems for a supercritical water cooled reactor [5]

Containment isolation valves can be check valves in feedwater lines, which need to be damped to avoid a water hammer, and steam isolation valves with hydraulic and medium controlled actuators as described by Sempell in [6]. A pressure suppression pool with vent pipes is keeping the containment pressure below the design limits, which can also serve as a heat sink for the automatic depressurization system in the simplest case. A low pressure coolant injection system with a heat exchanger to a secondary emergency coolant system is shown underneath the reactor in Fig. 4.2, but is meant to be placed somewhere inside or outside the containment in a sheltered position. These systems look quite similar as those of conventional BWR. The different response of these systems in the HPLWR, however, shall be discussed by studying a loss of feedwater accident.

Let us imagine the case of a simultaneous trip of all feedwater pumps caused e.g. by a station black out. These feedwater pumps are high pressure, multistage centrifugal pumps which must be equipped with a check valve each to avoid backflow in case of a trip of a single pump. These check valves, as well as those for containment isolation, will stop the feedwater flow within a few seconds and even a potential flywheel of the feedwater pumps could not extend the short coast down time. Different from a feedwater pump trip in a BWR, therefore, this case is equivalent with a loss of coolant flow to the core within a few seconds, requiring a scram of the reactor. As a consequence, the system must be depressurized immediately, being the only option to maintain a coolant mass flow rate through the core, either through the turbines and through the turbine by-pass valves as an immediate action, or through the automatic depressurization system inside the containment to avoid loss of coolant to the outside of the containment. It is not wise to close the turbine governor valve in this case to keep a high system pressure, like in a BWR, as such measure would stop the steam flow simultaneously, which would overheat the core.

The pressure and coolant temperature history in case of containment isolation of all feedwater and steam lines has been simulated by Schlagenhaufer et al. [7] for the HPLWR with its coolant flow path as described in Chapter 2.2. We see in Fig. 4.3 that the containment isolation will first cause a short pressure peak, which actuates the automatic depressurization system (ADS) of the steam lines, followed by rapid pressure decrease.

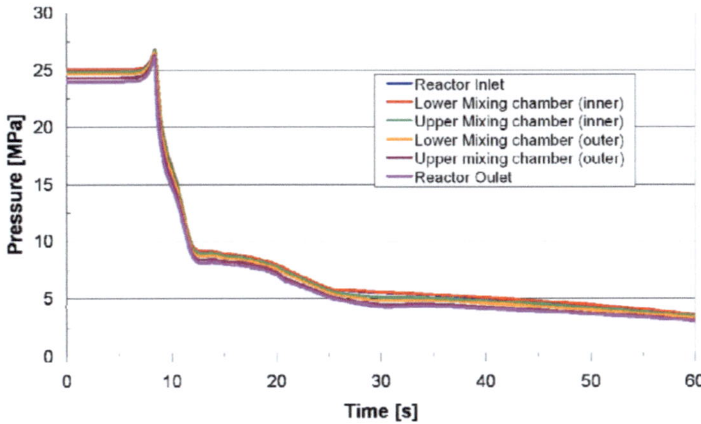

Fig. 4.3: Pressure history during a depressurization transient after containment isolation [7]

The temperature history shown in Fig. 4.4 gives a short temperature peak, caused by a 0.6 sec delay time of the ADS and by the 3.5 sec shut down time of the control rods, but the coolant temperature falls rapidly afterwards to the feedwater temperature of 280°C because of the high coolant mass flow rate during depressurization. Within 20 sec after containment isolation, the pressure has reached the saturation pressure of the feedwater inside the reactor and the feedwater in the upper plenum will start boiling. This situation will keep a minimum pressure in the vessel of initially 6.4 MPa which is decreasing slowly such that the core will be well cooled for about 10 min. If the low pressure coolant injection system will be designed with a pressure head of at least 6 MPa, and if the emergency power supply and the ramp up of the coolant injection pump can be provided within 20 sec in total, the core will be well cooled for long term as the cooling circuit is now closed within the containment before a significant amount of coolant was lost, and the residual heat will be removed to the secondary coolant system. This time is short but feasible in conventional boiling water reactors. It would be advantageous, however, to have a longer grace period.

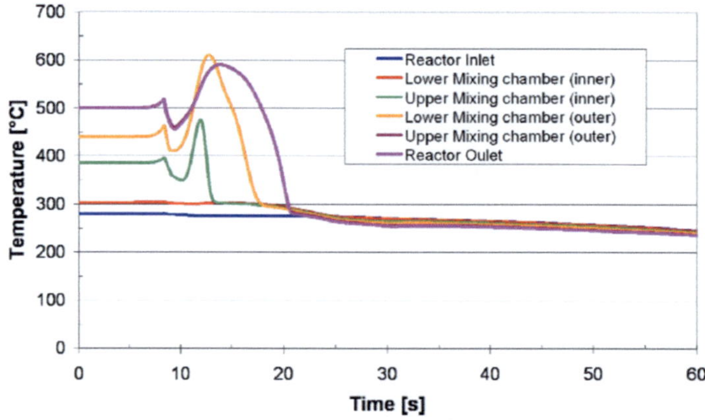

Fig. 4.4: Coolant temperature history during a depressurization transient after containment isolation [7]

The feedwater volume stored in the reactor pressure vessel has sometimes been called an "in-vessel accumulator" [8], suggesting that this water volume may be spent for cooling before the low pressure coolant injection becomes available. Indeed, the core is well cooled by this water for about 10 min, as described above. As soon as cold water is injected, however, the steam pressure in the reactor breaks down and the core flow is interrupted until the reactor has been filled up again. The in-vessel accumulator acts as a pressurizer, providing a driving pressure head for the coolant only as long as its temperature is sufficiently hot.

The problem can be overcome if the system is depressurized through a steam turbine driving a high pressure coolant injection pump, as sketched in Fig. 4.5. Such system has often been used already in conventional BWR. As the condenser behind this turbine must be at lower elevation than the turbine outlet, but the pump intake must be lower than the water reservoir, this concept is usually designed with 2 coolant pools at different elevation. As sketched in Fig. 4.5, Ishiwatari et al. [9] propose to use a separate condensate storage tank at lower elevation, like in a BWR. Now the missing coolant will be refilled already during depressurization. The steam mass flow must be high enough to ensure that the maximum cladding surface temperature in the core does not exceed the envisaged limit, but small enough to maximize the grace period for the active, low pressure coolant injection system.

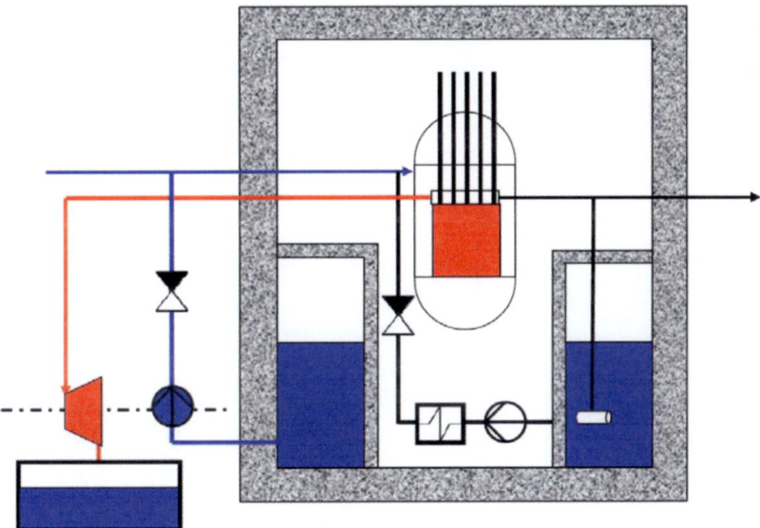

Fig. 4.5: Depressurization through a steam turbine driving a high pressure coolant injection pump [5]

A passive system without rotating components could be a closed loop which condenses the steam in an additional upper pool inside the containment, as sketched in Fig. 4.6. This system decreases the system pressure slowly, but the flow rate could eventually be too small to cool the core. Therefore, de Marsac et al. [10] propose to drive the coolant loop additionally with a steam injector. After an initial short depressurization through the ADS, the subcritical steam is supplied to the steam injector which drives a closed coolant loop through the condenser in the upper pool. Coolant is lost to the containment pool only during the short initial depressurization phase, and the steam supplied to the injector afterwards is condensing in the closed system. This innovative system, however, has never been analyzed in detail yet.

Instead, Schlagenhaufer et al [7] propose to use a motor driven recirculation pump to drive the closed coolant loop through the condenser in the upper pool, which is easier to control and thus easier to optimize for this purpose. Figs. 4.7 (a) to (d) show the response of the reactor coolant during a depressurization transient using this system. The transient was initiated at time step 5 sec by inadvertent containment isolation, which caused a pressure peak, scram and activation of the ADS as described above. Despite the peak mass flow rate of the steam leaving the reactor outlet (a), a short temperature peak of the coolant cannot be avoided again (b). Simultaneous with scram activation, a recirculation pump in the condensate line of the closed loop is started now, and the ADS is closed again as soon as the pressure is less than 10 MPa. As a consequence, the coolant temperature at reactor inlet drops suddenly to 20°C at time step 20 sec, as the condensate stored in the loop has been cold during normal operation.

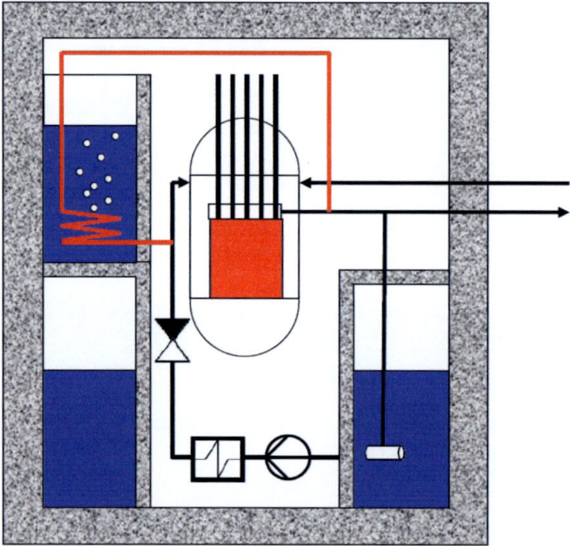

Fig. 4.6. Depressurization by condensation in a closed loop inside an upper containment pool [5]

Around 1 min after scram, the closed loop has stabilized, and the condensate temperature increased to the saturation temperature at actual system pressure (c). The core outlet temperature is controlled by the recirculation pump such that is stays slightly superheated, which minimizes the coolant mass flow rate and thus the required power of the pump. Fig. 4.7 (d) shows more than 90% void in the core after 5 min and even the feedwater inside the reactor "in-vessel accumulator" is boiling, but the core remains to be cooled sufficiently. The total peak power of 4 recirculation pumps needed for this system was 1 MW.

This exercise could serve as a starting point for a passive system, e.g. with a condenser at higher elevation. Fully passive safety systems, however, which could remove the residual heat over several days without any auxiliary power supply, have not been designed yet, and it is not even clear up to now if they can ever be designed with reasonable effort. A general difficulty of the HPLWR core design is its rather high coolant pressure drop as well as the meandering coolant flow path in the core which require a higher pressure head of a passive coolant loop than for a BWR. Therefore, analyses of the safety systems, which have been performed exemplarily up to now, were rather concentrated on active residual heat removal systems. Nevertheless, several passive systems were already included in the HPLWR design studies to prepare a basis for future optimization, as will be discussed next.

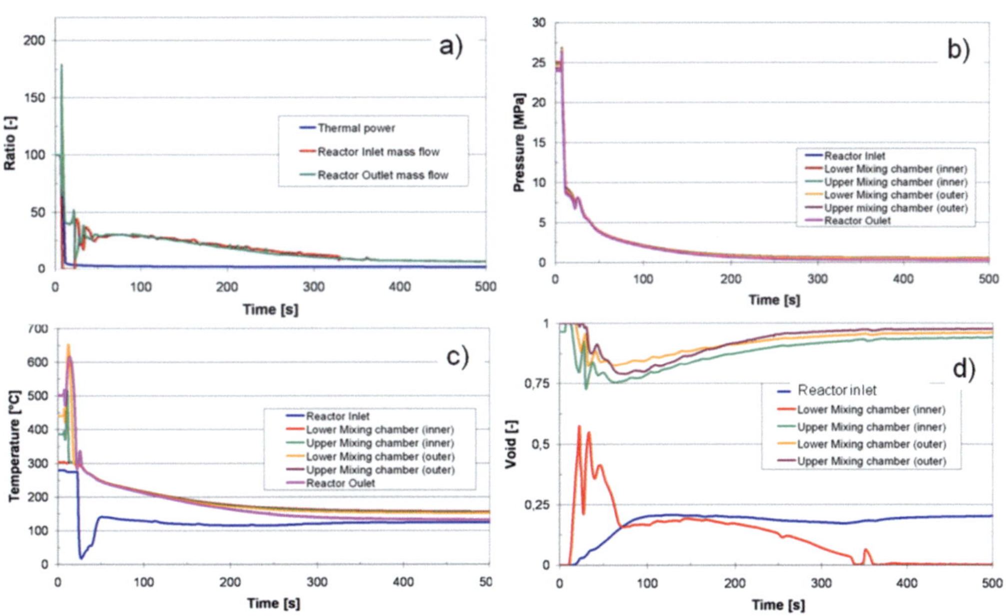

Fig. 4.7: Depressurization transient using a closed loop with condenser in an upper pool inside the containment [7]; (a) core power and coolant mass flow rate, (b) coolant pressure, (c) coolant temperature, (d) coolant void fraction

4.2   Design of the containment and its safety systems

A containment with safety systems as sketched in Fig. 4.6 is the compact HPLWR containment shown in Fig. 4.8 with 20 m inner diameter and 23.5 m inner height. The cylindrical containment from pre-stressed concrete is designed for an internal pressure of 0.5 MPa. It contains the reactor pressure vessel, an annular pressure suppression pool with 900 m^3 water and 500 m^3 nitrogen, 4 upper pools with a total water volume of 1121 m^3, and a drywell gas volume of 2131 m^3. Four feedwater lines with check valves and four steam lines with containment isolation valves, each inside and outside of the containment, connect the reactor with the steam cycle. They are assumed to have a stroke time of 3 sec, closing both actively as well as passively. Four automatic depressurization systems, each equipped with 2 safety relief valves and 2 depressurization valves, open a flow cross section of 110 cm^2 each to 8 spargers in the upper pools. The actuation pressure of the safety relief valves has been set to 27.5 MPa.

Underneath the pressure suppression pool, 4 redundant and separated low pressure coolant injection pumps with an outlet pressure of at least 6 MPa and a maximum flow rate of 180 kg/s each supply coolant from the pressure suppression pool via a heat exchanger for residual heat removal and via a check valve to the feedwater line. Overflow pipes from the upper pools to the pressure suppression pool close the coolant loop inside the containment. 16 vent tubes for pressure suppression in the containment connect the drywell with the pressure suppression pool.

Four emergency condensers are connected with the 4 steam lines and with the 4 feedwater lines, hanging from top in the upper pools. Flow through these condensers is driven by a steam injector as will be described below. In addition, there are 4 containment condensers mounted at the ceiling of the drywell which are connected on their secondary side to pools above the containment. The secondary side is permanently open so that steam in the containment can condense as soon as the saturation temperature in the pools has been reached and the containment pressure is starting to rise, in the unlikely case that the heat sink of the residual heat removal system were not available. Open connecting pipes from the ceiling to the pressure suppression pools enable a discharge of hydrogen from the drywell. The pressure suppression pool, in turn, can be vented to the stack through aerosol and iodine filters.

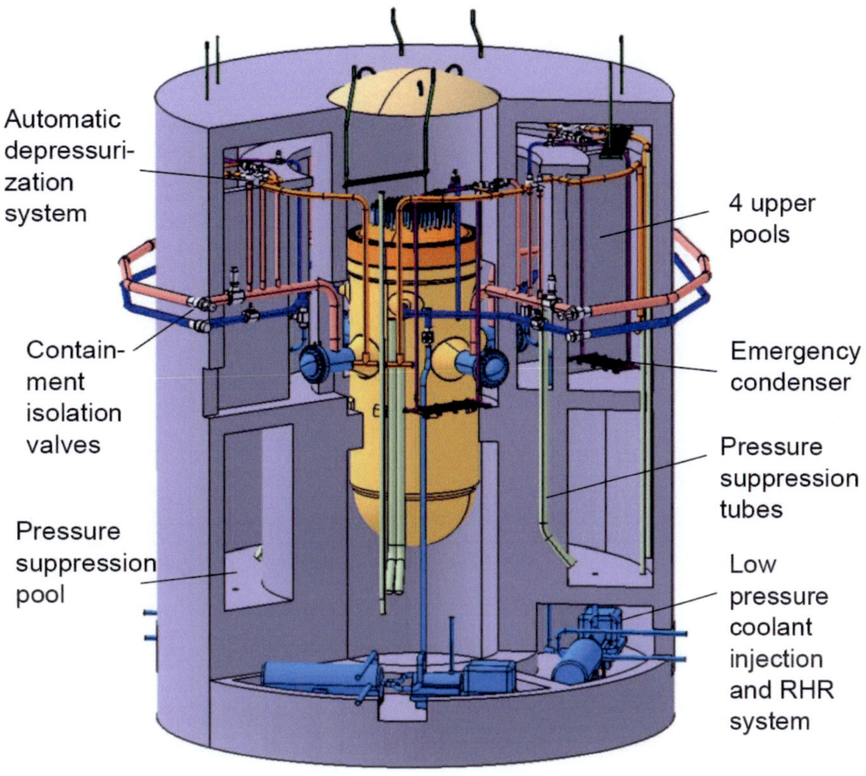

Fig. 4.8: HPLWR containment with active and passive safety systems

Outside the containment, a boron poisoning system on top of the containment with a tank of about 10 m^3 of B-10 with a concentration of 20 to 25% is connected with the feedwater lines by 2 lines including pumps. This system is not included in Fig. 4.8. It serves as the second, redundant shut down system.

The emergency condensers in the upper pools are shown in more detail in Fig. 4.9. This closed loop residual heat removal system has been proposed by de Marsac et al. [10] based on simplified analyses. It consists of 4 steam injectors which are connected with the steam lines, driving a condensate flow through a closed loop in the upper pool. Flow through the steam injector is initiated by opening a bleed valve to the spargers of the depressurization lines. As soon as sub-cooled condensate from the cooler in the upper pool is sucked into the steam injector, the steam jet condenses inside the steam injector, which builds up a condensate pressure opening the check valve to the feedwater line. Once the flow has been established, the bleed valve is closed. Now the pressure vessel is depressurized slowly through the steam injectors, which build up enough pressure to refill condensate into the vessel and to drive a coolant flow through the core. The coolant mass flow is controlled by the control valve in the condensate loop, shown in Fig. 4.9, such that the core outlet temperature remains to be slightly superheated. This system does not need any power to drive a pump, but auxiliary power for the control system driving the valves is still needed.

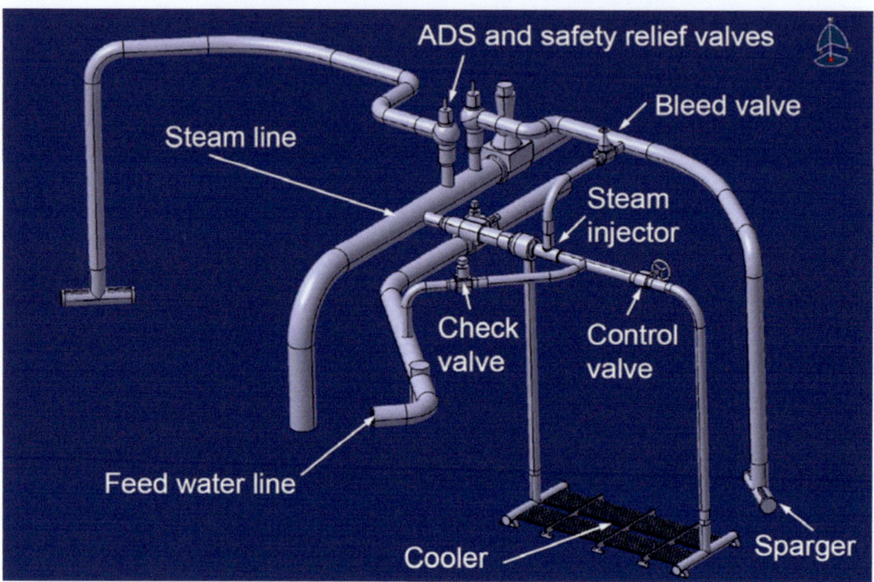

Fig. 4.9: Closed loop steam condensation driven by a steam injector [10]

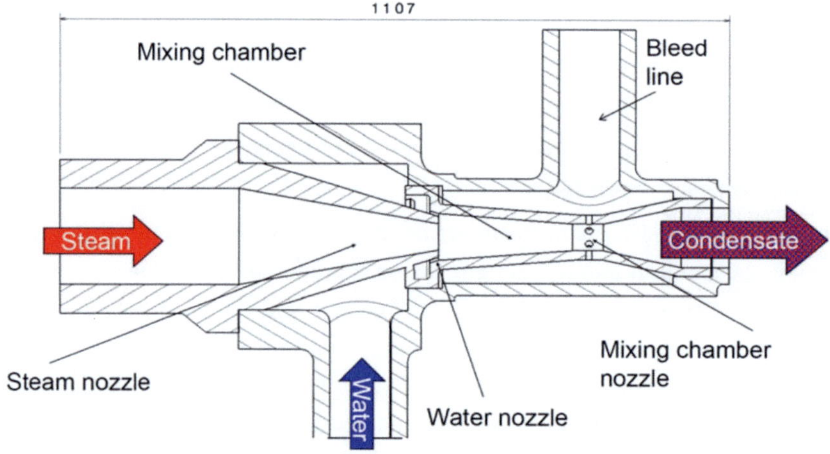

Fig. 4.10: Conceptual design of the steam injector

A conceptual design of the steam injector is sketched in Fig. 4.10. Such steam injectors had often been used in locomotives many years ago. More recently, a steam injector as a passive thermal-hydraulic component for pressurized water reactors has been tested by Dumaz et al. [10]. They could demonstrate that an emergency feedwater system for a VVER would refill the steam regenerator successfully over a period of more than 1 hour. Compared with Fig. 4.10, their steam injector had been improved by a control needle inside the steam nozzle and by a Laval nozzle providing supersonic steam velocities.

The balance equations for mass, energy and momentum, applied between the outlets of steam and water nozzles and the condensate nozzle, help to estimate the performance of the steam injector. The mass balance can be written as

$$\dot{m}_S + \dot{m}_W = \dot{m}_C \tag{4.1}$$

Where $\dot{m}_S$, $\dot{m}_W$ and $\dot{m}_C$ denote the mass flows of steam, water and condensate, respectively.

Similarly, the energy balance reads

$$h_S\,\dot{m}_S + h_W\,\dot{m}_W = h_C\,\dot{m}_C \tag{4.2}$$

Where h_S, h_w and h_C denotes the enthalpies of steam, water and condensate, each at rest.

The contact of superheated steam with the sub-cooled water causes the steam pressure to drop to the saturation pressure of the sub-cooled liquid, which drives a sonic flow in the steam nozzle. The pressure in the mixing chamber rises afterwards again, first to the saturation pressure of the mixture, followed then by a condensation shock in the condensate nozzle with a sharp increase of pressure. Taking the mixing chamber as a black box, we can formulate the momentum balance between the steam and water nozzles on the left and the condensate nozzle at the right as

$$p_S\,A_S + \dot{m}_S u_S + p_W\,A_W + \dot{m}_W u_W = \left(A_S + A_W\right)p_C + \dot{m}_C u_C \tag{4.3}$$

Here, p_S and u_S denote the steam pressure and velocity in the steam nozzle, which can be derived from the steam conditions at rest by assuming an adiabatic expansion to sonic velocity, p_W and u_W are the water pressure and velocity in the water nozzle, which can be estimated by Bernoulli's equation, and p_C and u_S are the velocities in the condensate nozzle, assuming that the steam has fully been condensed before. The velocity in the condensate nozzle can be written as

$$u_C = \frac{\dot{m}_C}{\rho_C A_C} \tag{4.4}$$

A_S, A_W and A_C denote the cross sections of the steam nozzle, the water nozzle and the condensate nozzle, respectively, and ρ_C denotes the condensate density, which is a function of the condensate enthalpy. The condensate pressure can further be increased by a diffuser as sketched in Fig. 4.10.

The system of equations is closed by considering on one hand the pressure losses and heat transfer through the control valve and the cooler, Fig. 4.9, and on the other hand the pressure losses and heat input through the reactor. De Marsac [12] solved these equations for quasi steady state operation with Excel, assuming a pool temperature of 100°C and a residual heat input P according to

$$P = 0.0622\,P_{th}\,/\,t^{0.2} \tag{4.5}$$

Where P_{th} is the thermal power of 2273 MW before scram. The coolant loop was assumed to be controlled such that the steam will constantly be superheated to 400°C. Results are shown in Fig. 4.11.

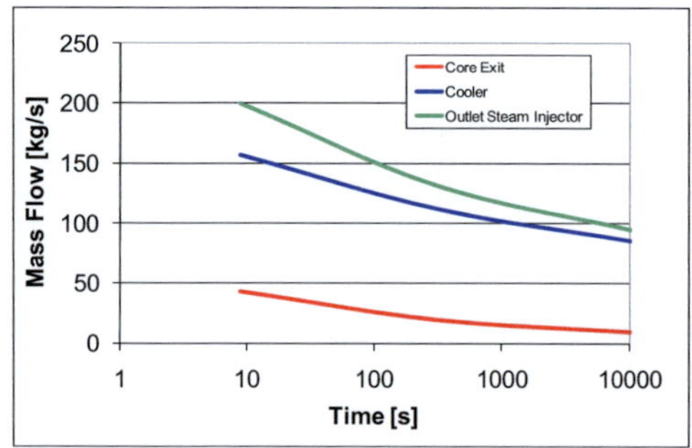

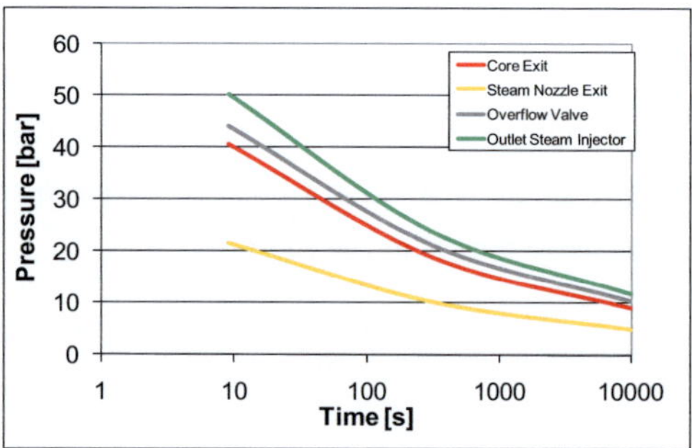

Fig. 4.11: Quasi steady state analysis of pressure and mass flows provided by an ideal steam injector [12]

In reality, this predicted pressure increase will be difficult to achieve, as shown by the experiment of Dumaz et al. [10], and a more detailed analysis as well as an experimental verification will be needed to confirm these results.

4.3   System response to postulated accidents

The transient system response has been analyzed with the system code APROS by Schlagenhaufer [13] and with CATHARE, RELAP5, ATHLET-KIKO3D and SMABRE-TRAB-3D as summarized by Andreani et al. [14].

Modeling of supercritical water with system codes does not only require adding the properties of supercritical water as a single phase fluid to the list of coolants. If the transition from supercritical pressure to subcritical pressure causes the critical point of water to appear somewhere in the fluid domain, the code must differ locally between two-phase and single phase flow. The codes RELAP5 and ATHLET are still suffering from numerical problems at this stage.

An approach to overcome these problems has been proposed by Antoni and Dumaz [16] who modeled supercritical water as a pseudo two-phase flow with a six-equation model, which is physically equivalent to the single phase supercritical fluid, but numerically structured as a two phase medium. The improved model treats the single-phase supercritical fluid as liquid if its enthalpy is below the pseudo-critical line and as gas otherwise. These states are referred to as pseudo-liquid and pseudo-gas to emphasis the fact that at supercritical pressures there are no distinct phases. Later, the separate two-fluid model of APROS has been upgraded similarly by Hänninen and Kurki [19] to cope with the supercritical-pressure conditions.

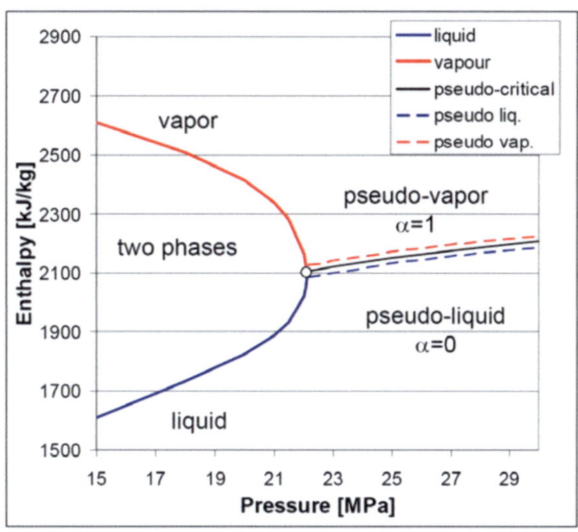

Fig. 4.12: Modeling of supercritical water with APROS and CATHARE [5]

The method is illustrated in Fig. 4.12. If the fluid enthalpy is less than the pseudo-critical enthalpy, the fluid is treated as liquid with zero void α, and as vapor with $\alpha=1$ if the enthalpy is greater than the pseudo-critical enthalpy. A small artificial evaporation enthalpy is introduced to model the continuous transition as a small step when passing the pseudo-critical line. The flow is modeled with 6 equations for conservation of mass, momentum and energy, assuming that both pseudo-phases have the same velocity, that stratification is zero, and that surface tension is zero, which avoids any liquid entrainment. In the narrow range of pseudo-vaporization, the void is changing from zero to one like with sub-critical fluids. This approach allows modeling the supercritical range with the same set of equations as sub-critical fluids with a change of the parameters only. The additional uncertainty of pseudo-evaporation can be minimized by reducing the artificial heat of vaporization. Both codes, CATHARE and APORS, worked successfully with a smooth transition to sub-critical pressures in all transient analyses.

A number of heat transfer correlations and a pressure drop correlation developed specifically for the supercritical pressure region have been incorporated. The thermo-physical properties of water (i.e. the steam tables) of APROS have been extended and refined in order to describe the supercritical pressure region with adequate accuracy. The steam tables are based on the IAPWS-IF97 recommendation.

Schlagenhaufer [13] used APROS to model each heat up step of the core as a common channel with integral properties and a small, parallel channel, simulating a hot channel as discussed in Chapter 2.12 with twice the enthalpy rise under steady state conditions. A neutronic feedback had not yet been included in his model. The nodalization of a simplified three-pass-core is shown in Fig. 4.13. The reactor core was simplified by standard heat transfer components which have 10 nodes in axial direction each. The power distribution shown in Fig. 2.15 was assumed for the APROS core model. The generated power in the evaporator (EVA), superheater 1 (SH1) and superheater 2 (SH2) was 48%, 35% and 17% of the thermal power.

The safety systems with the necessary intersections to the containment were included in the APROS steam cycle model as described by Schlagenhaufer [13]. However some simplifications had to be made to decrease complexity and calculation time to an acceptable limit. The APROS containment model contained:

- the containment isolation valves, which are the main feedwater isolation valves (MFIV) and the main steam isolation valves (MSIV)
- the feedwater and steam lines
- a single upper pool combining the 4 pools shown in Fig. 4.8.
- the pressure suppression pool (PSP)

- the automatic depressurization system (ADS)

- the active low pressure coolant injection (LPCI) system and

- a closed loop, high pressure coolant injection (HPCI) system with a condenser in the upper pool and a motor driven recirculation pump.

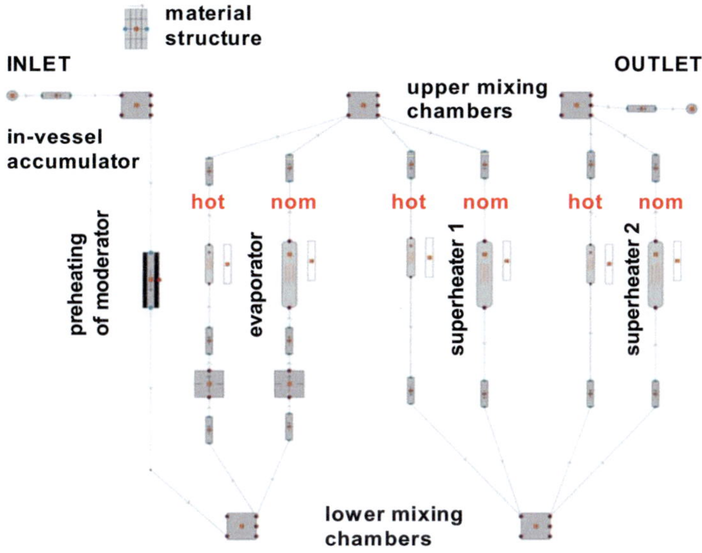

Figure 4.13: Nodalization of the three pass design concept of the HPLWR core [13]

The LPCI system was a four time redundant system, which consisted of a heat exchangers and a pump, as shown in Fig. 4.8. In the APROS containment model, the heat exchanger was omitted since the heat-up of the PSP was neglected and only a single pump was modeled. The LPCI pump was started after the reactor inlet pressure dropped below 6 MPa. The reactor pressure was measured with a delay time of 0.1s. The rotation speed was assumed to increase from 0 to 100% within 1s after the LPCI pump started and assumed to stay constant for the rest of the transient. The LPCI pump parameters are presented in Table 4.2.

The condenser of the HPCI system, as sketched in Fig. 4.6, was dimensioned to treat 200 kg/s of superheated steam (10 MPa; 500 °C) and to cool it 20 °C below the saturation temperature. The ADS shut valves had a driving time of 2s and a total flow area of 0.1608m^2 like the ADS piping. Table 4.2 depicts the main input parameters for the HPCI system.

	Nominal mass flow rate [kg/s]	Nominal head [m]	Maximal head [m]	Suction head [m]
LPCI pump	400	602	752	7.18
HPCI pump	200	42	52	9.3
	Hydraulic diameter [m]	Wall thickness [m]	Pipe length [m]	Number of pipes [-]
HPCI condenser	0.015	0.001	3	7000

Table 4.2: Main input parameters for LPCI and HPCI system

Additionally needed parameters like the actuation pressure, the driving time and the flow area of the ADS valves and the driving time of the containment isolation valves were varied as will be described next. A parametric study has been performed in [13] to evaluate the influence on the hot channel cladding temperatures and the peak pressure at reactor inlet, since they were identified as the most crucial parameters. The goal of this parametric study was to identify settings to be recommended for further analyses of the safety system. The varied system parameters for the depressurization analyses were:

I.   Actuation pressure of ADS-system

II. Driving time of ADS valves

III. Driving time of main feedwater isolation (MFIV) and main steam isolation valve (MSIV) and

IV. Flow area of ADS valves.

The study included 4 cases, in which one of these four parameters was varied and the other three ones were kept constant. As an example, the first one of the parametric studies is summarized here. The complete analysis is reported in [13].
The event history for the parametric study is depicted in Fig. 4.14.

An inadvertent isolation of all MFIV and MSIV was assumed to occur after 5s of normal operation. The ADS valves open in 0.2s after the pressure in front of the ADS valves rises above the actuation pressure of 26 MPa. A signal with a delay time of 0.6s is sent to the reactor SCRAM system, the control rods are inserted into the core and the thermal power is reduced linearly to 6.22% within 3.5s. After that, the thermal power follows the decay heat function, eq. 4.5.

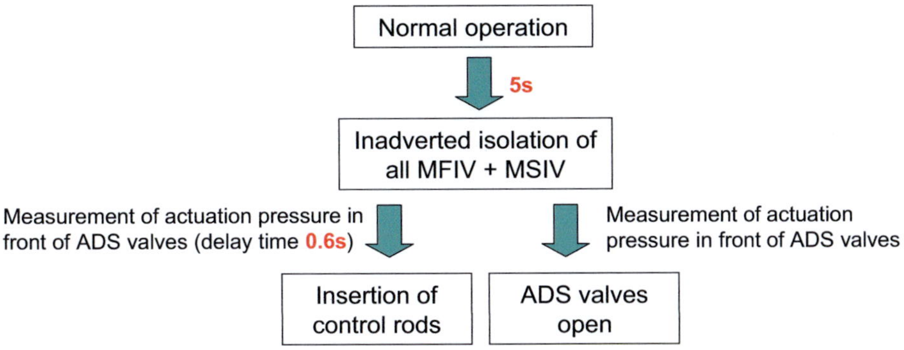

Figure 4.14: Event history for depressurization transients

The system parameters, which have to be defined for the parametric study I, are described in Table 4.3. The actuation pressure of the ADS valves was varied between 24.5MPa and 29 MPa in steps of 0.5MPa, whereas the MFIV, the MSIV and ADS valves had a driving time of 3s, 3s and 0.2s respectively. The total flow area of the ADS valves was $0.09m^2$.

Driving time of MFIV and MSIV [s]	Total flow area of ADS valves [m^2]	Driving time of ADS valves [s]	Actuation pressure of ADS-system (varied) [MPa]
3	0.09	0.2	24.5 - 29

Table 4.3: Input parameters for parametric study I

The peak of the maximum hot channel cladding temperatures and the peak reactor inlet pressure due are shown in Fig. 4.15 as a function of the ADS actuation pressure. The peak reactor inlet pressure increases linearly with increasing ADS actuation pressure. This is obvious, since the reactor pressure can increase until the ADS opens at a certain pressure and the depressurization starts. But attention should be paid that the maximum occurring reactor pressure must not exceed the design pressure of the reactor pressure vessel (28.75 MPa), which means that the actuation pressure should not be higher than 27.5MPa. The peak hot channel cladding temperatures show a similar behavior, which decrease with lower ADS actuation pressure.

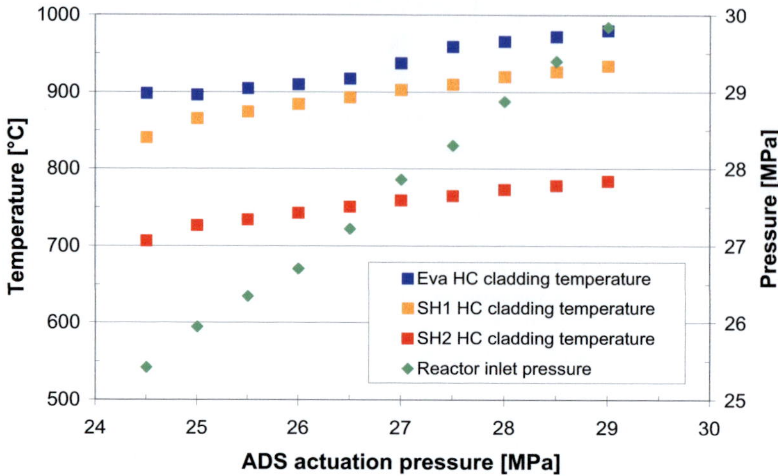

Figure 4.15: Peak hot channel cladding temperatures and peak reactor inlet pressure due to ADS actuation pressure variation [13]

Based on this and on the other parametric studies mentioned above, the main settings for the ADS system of the HPLWR, as listed in Table 4.4, can be recommended for further analyses.

Driving time of MFIV and MSIV [s]	Total flow area of ADS valves [m2]	Driving time of ADS valves [s]	Actuation pressure of ADS-system [MPa]
3	0.09	0.2	26

Table 4.4: Input parameters for ADS valves and spargers

The simulation of a depressurization transient showed that the ADS system with the chosen parameters was able to limit the cladding temperature excursion in the short-term. For the long-term cooling, the effect of the LPCI in combination with the ADS actuation was analyzed.

4.3.1 Simulation of the main safety LPCI system

The LPCI system was simulated with the pump parameters presented in Table 4.2.

The event history for the depressurization with start-up of the LPCI system is sketched in Fig. 4.16. An inadvertent isolation of all MFIV and MSIV was assumed to occur after 5s of normal operation (classified as a DBC3 event). The ADS valves open in 0.2s after the pressure in front of the ADS valves rises above the actuation pressure of 26 MPa. A signal with a delay time of 0.6s is sent to the reactor SCRAM system and the control rods are inserted into the core within 3.5s. Another signal is sent to the LPCI pump, which is started within 1s, if the pressure at reactor inlet falls below 6MPa. For the measurement of the reactor inlet pressure, a delay time of 0.1s is assumed.

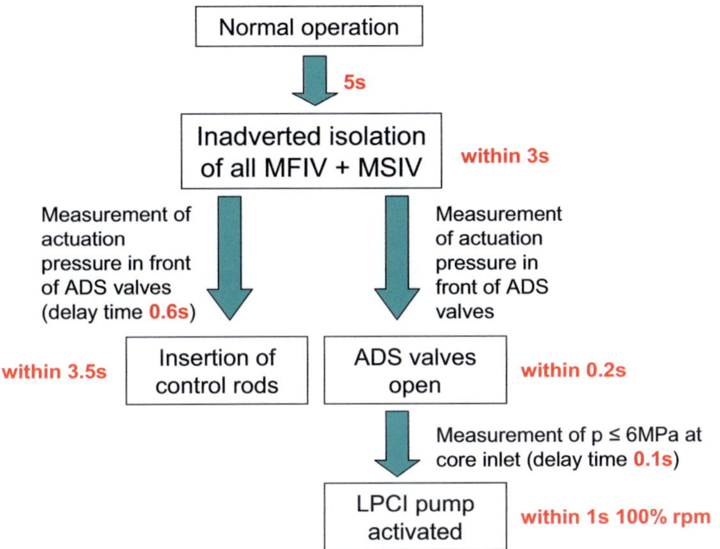

Figure 4.16. Event history for depressurization and start-up of the LPCI system

The cladding temperatures during the depressurization transient with start-up of the LPCI pump are shown in Fig. 4.17. The peak cladding temperatures of nominal channels are not of concern, but the peak cladding temperature of the evaporator hot channel rise by about 370°C for almost 10s, whereas a smaller increase in other hot channels can be observed.

A region with almost constant cladding temperatures in all core regions can be observed after 70s.

Figure 4.18 shows the void fraction at several locations in the core and above it (called the water accumulator). It can be observed that the core is completely refilled with liquid within about 140s.

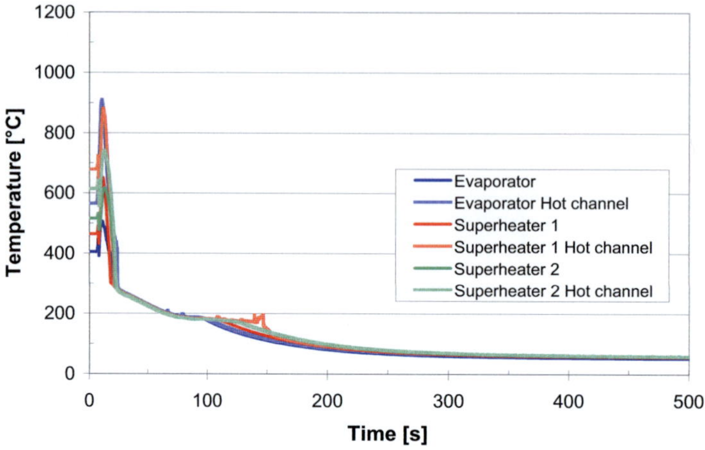

Figure 4.17: Cladding temperature of nominal and hot channels during LPCI [13]

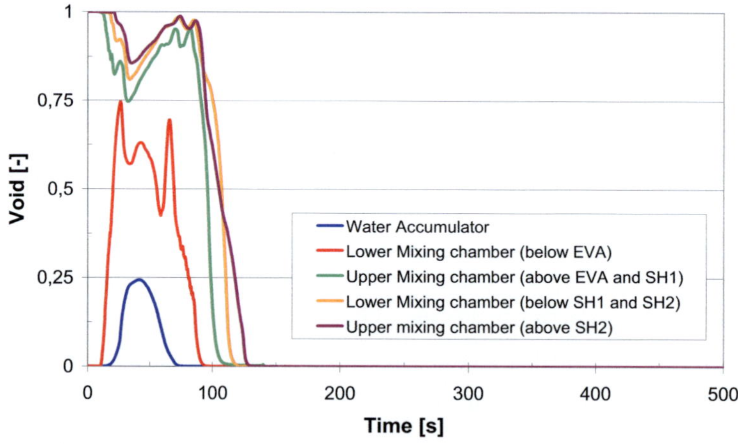

Figure 4.18: Void of reactor volumes during LPCI [13]

The maximum electric pump power was calculated to be 2950kW for the LPCI system, which can be delivered by a diesel generator in case of a station black-out.

The simulation of complete containment isolation with following depressurization through the ADS system showed that the reactor can be cooled efficiently. The ADS actuation pressure, the ADS valves flow area, the ADS valves driving time and the MFIV and MSIV driving time were varied. These studies indicated that an ADS actuation pressure of 26MPa, a flow area of $0.09m^2$ and a driving time of 0.2s are an optimal set of parameters.

The simulation of the LPCI system showed no principle drawbacks. The reactor can be cooled efficiently after the system is depressurized through the spargers. After the LPCI system injects water at 6MPa, the cladding temperature starts to rise again, since the reactor has to be filled with water. The simulation with a pump with 400kg/s nominal mass flow rate prevents almost the rising of the cladding temperature, since the core remained to be flooded. But it has to be taken into account that the pump run with full speed, which means that the effective injected mass flow rate increases due to the falling backpressure of the system.

4.3.2 Transients addressing heat storage capacity

Heat storage capacity within the primary system is considerably less compared to a PWR and BWR, which is an indication for potential of faster pressure transients. In order to address this issue, an accident was assumed in which the turbine was tripped, but the turbine bypass valve did not open, to check if the ADS could limit the pressure below the design pressure (classified as a DBC4 event). Two analyses were performed with two codes, using different models and assumptions:

- RELAP5: point kinetics were used. No scram was assumed, and the calculation showed that power was reduced to a new equilibrium value due to the negative reactivity effect. Since the RELAP5 cannot simulate depressurisation below the critical pressure, the actuation of the ADS system was not simulated. Instead, the safety valves opened on high-pressure signal and cycled around the opening set point (26 MPa). Three different cases were investigated (Table 4.5), using different assumptions on MSIV valve closing time and safety relief valve (SRV) opening/closing pressure.

- SMABRE/TRAB-3D: A 3-D Thermal-Hydraulic analysis was coupled with neutronic analyses. The ADS system was simulated to open when the pressure reaches 26 MPa. Once opened, the valves remain open and the system depressurises. Due to the numerical problems of SMABRE for fast depressurization transients, only the first few seconds could be simulated. The effect of the total valve flow area has also been studied.

Case	MSIV closing time(s)	SRV opening/closing time (s)
1	0	0
2	1	0
3	1	0.5

Table 4.5: Cases considered for the analysis of a turbine trip transient

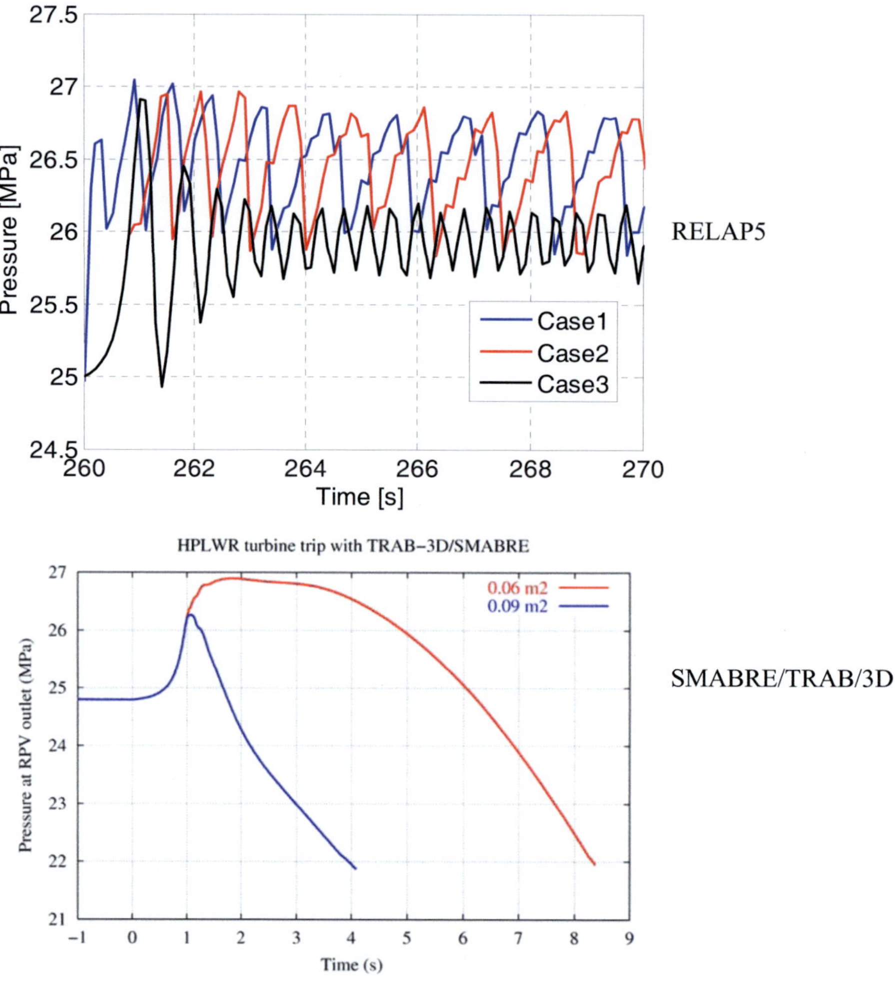

Figure 4.19: Pressure transient following a turbine trip without opening of the by-pass value, calculated with the codes SMABRE/TRAB-3D (top) and RELAP5 (bottom) [14]

A case study with RELAP (Figure 4.19, lower picture) has been carried out for opening and closing safety valves. Such a valve operation would keep the fluid in supercritical state, because the safety relief valve (SRV) closes before the transition to subcritical conditions might be reached. For three different cases given in Table 4.5, the pressure in the reactor pressure vessel is given in Figure 4.19. For all three cases, the maximum pressure is about 27MPa, which is beneath the design limit. In long term, case 3 shows the best performance because of pressure fluctuations with the smallest amplitude. Such an operation of the safety relief valves seems to be possible for the HPLWR.

4.3.3 Transients addressing core cooling in case of loss of flow

A variety of studies have been performed with the different system codes for sequences initiated by the failure of one or more pumps, as described in [15]. These simulations included the investigation of the effect of pump run-down times and of the parameters characterizing the intervention of the stand-by pump (delay and time for its actuation). Two representative studies are shown here to illustrate the most important conclusions of those analyses.

Partial loss of feedwater (LOFW):

Two of three feedwater pumps were assumed to fail, and a stand-by feedwater pump was assumed to intervene within a very short time (classified as a DBC2 event). The assumed time history of the total flow rate entering the vessel is shown is shown in Fig. 4.20. For this case, it was assumed that the feedwater flow rate of the 2 pumps which failed would run to zero in 10 s.

This transient has been analysed with CATHARE, RELAP5 and SMABRE, respectively. The main features of the models are listed below:

- SMABRE: the core was represented with a single channel (average) for each of the three core sections (Evaporator, Superheater 1 and Superheater 2). A hot channel has not been included. Each fuel channel was thermally connected with a moderator and a gap channel. The power distribution among the three regions of the core and its axial distribution were considered like in the APROS model described above.

- RELAP5: similar representation as with SMABRE, but the peak power in the hot channel was represented by modeling a second fuel rod in the average hydraulic channel, applying individual hot channel factors.

- CATHARE: the average assembly and the hot assembly are represented separately, resulting in the most detailed core model.

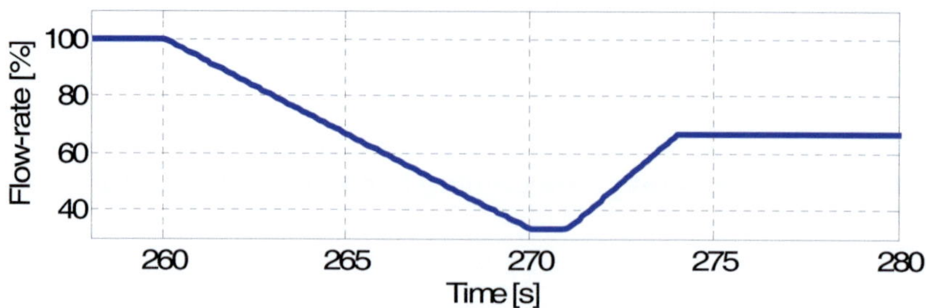

Figure 4.20: Total flow rate at the vessel inlet during the transient initiated by the loss of two FW pumps

For these studies, the assumption was made that the pressure was kept constant by control of the turbine governor valve.

Representative results for the peak cladding temperature are given in Figs. 4.21 and 4.22, where the calculation with the RELAP5 and CATHARE codes are shown for comparison. The cladding temperature excursion is rather small (around 50 K), due to the fast reduction in thermal power caused by the negative reactivity coefficient.

These results have to be taken with some caution because the use of point kinetics could be inadequate for this transient. The change in the flow distribution and even flow reversal in the gap channels, as shown in Fig. 4.23, results in variations of the moderator density distribution and therefore in modifications of the power distribution, both radial and axial.

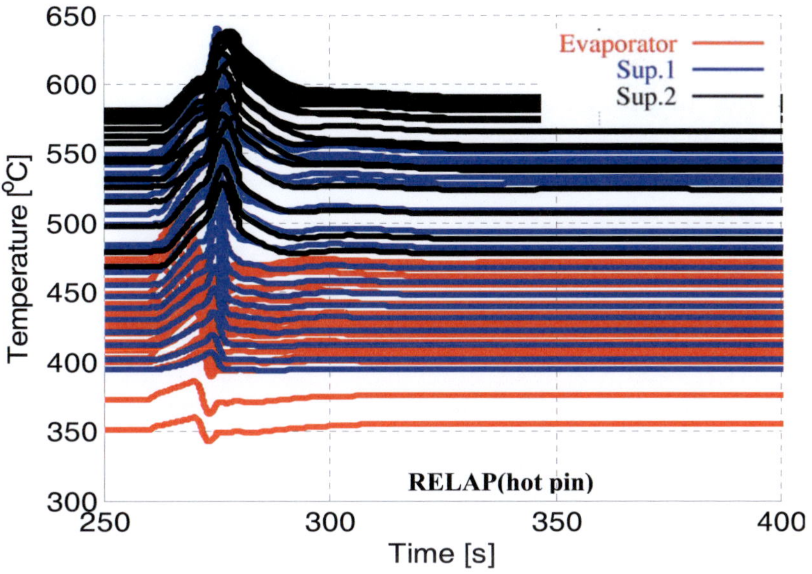

Fig. 4.21: RELAP5 analysis of peak cladding temperatures assuming the loss of 2 of 3 feedwater pumps and replacement by a hot stand-by pump without scram [14]

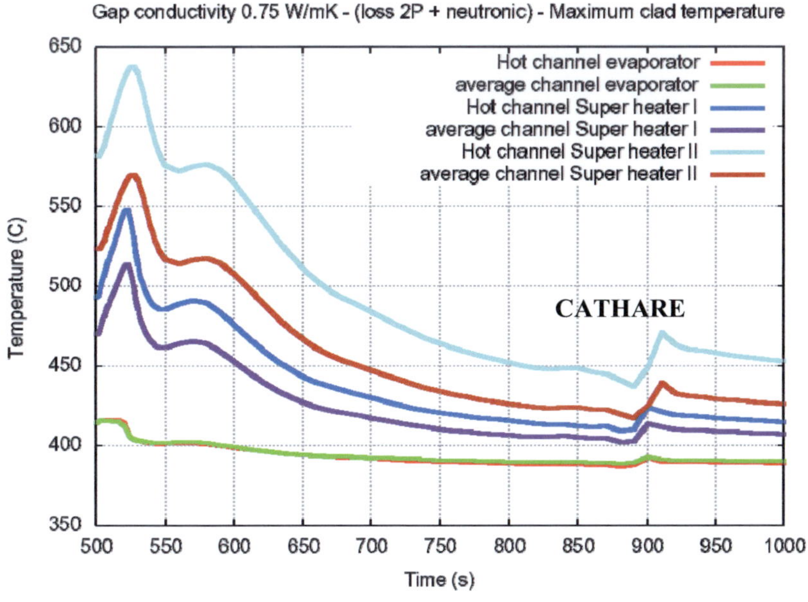

Fig. 4.22: CATHARE analysis of peak cladding temperatures assuming the loss of 2 of 3 feedwater pumps and replacement by a hot stand-by pump without scram [14]

During early analyses, reverse flow in the gap and moderator channels had been observed in several analyses when the core flow rate drops below 40% of the nominal value. To avoid this condition, the moderator flow path has been optimized to an upward flow in the gap volume between the assembly boxes, as explained in chapter 3.3.2. Moreover, it has been shown that the heat transfer from the fuel to the moderator channels boxes plays an important role, and therefore the use of a better insulating material for the moderator boxes would be beneficial.

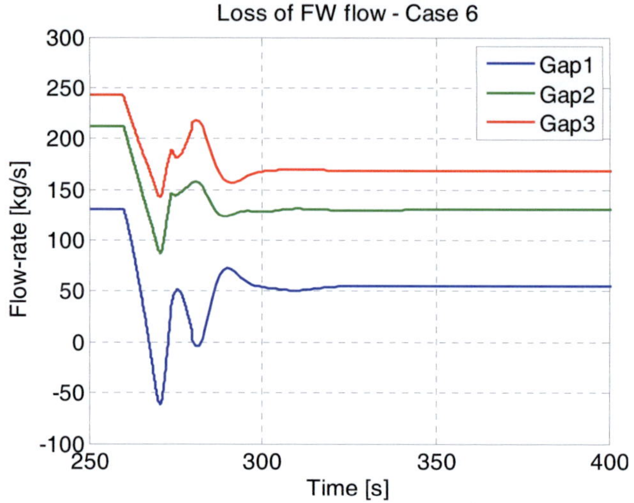

Figure 4.23: Flow in the gap channels calculated with RELAP5 for a loss of feedwater flow with failure of two pumps [15]

Total loss of feedwater

The extreme event of a total loss of feedwater was investigated under different conditions with three codes:

- APROS: a model with simplified representation of the core and a complete representation of the steam cycle was used by Schlagenhaufer [13]. The pressure evolved according to the response of the system, which included actuation of the ADS system and associated fast depressurisation.

- SMABRE: a detailed model of the core was used, and a pressure boundary condition was prescribed at the vessel inlet. The total feedwater flow rate was ramped to zero within 10s.

- RELAP5: the model had similar features as that which are used for SMABRE, the only important difference being that the hot channel was also represented.

For the simulation with APROS (which does not model reactivity feedback), a reasonable power time history after scram was imposed (DBC2). Calculations with SMABRE and

RELAP5, however, assumed that scram would not occur (the transient being therefore a very severe ATWS, classified as DBC4).

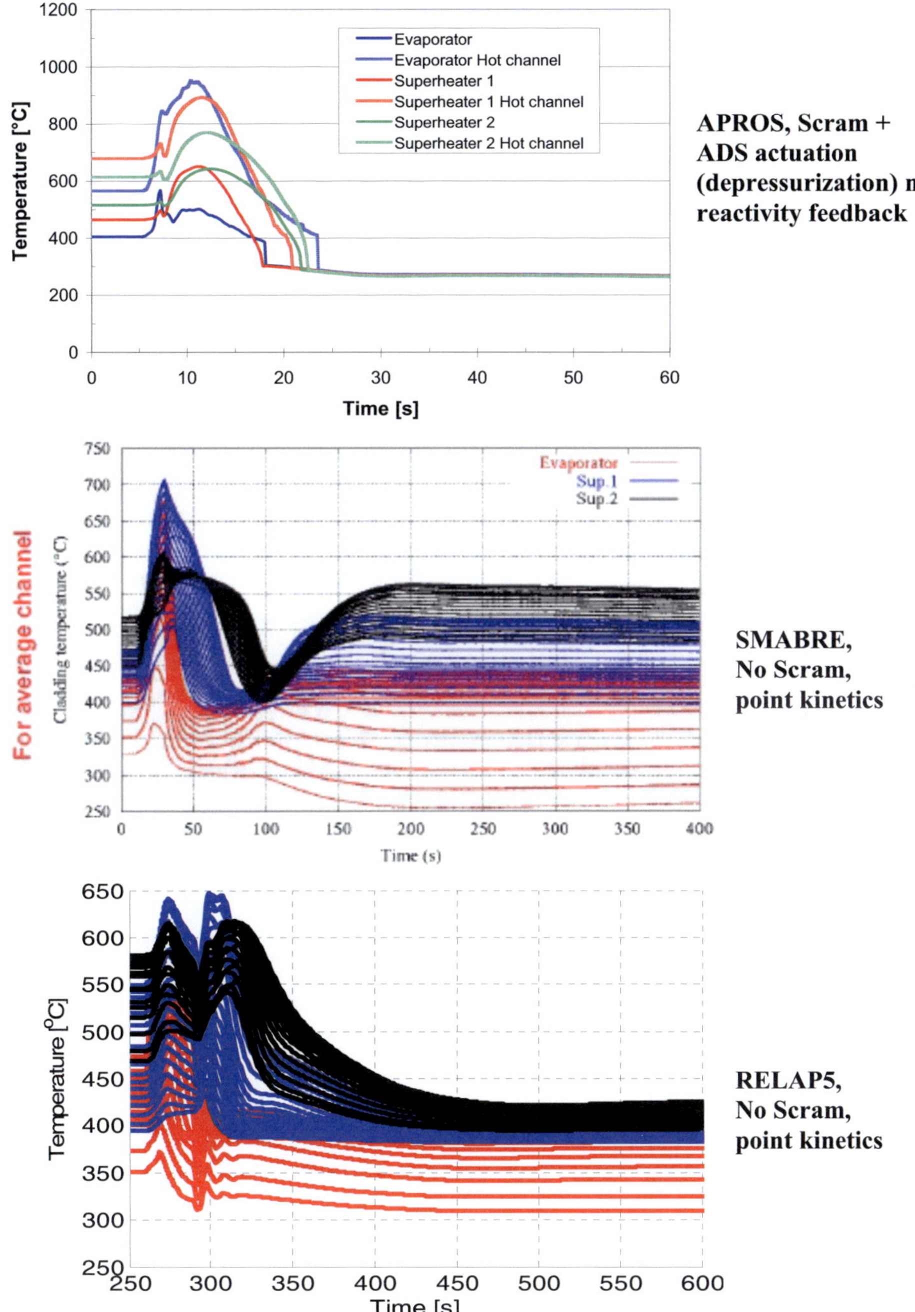

Figure 4.24: Peak cladding temperatures for a total LOFW calculated with three codes [15]

In the models for these two codes, different temperatures were adopted for the auxiliary flow, which is started at 20 s. Moreover, the way to account for the hot channel in the APROS model was different from the method used in the simulation with RELAP5: in the APROS model, a hot channel was represented, assuming that the fluid enthalpy rise is twice as that in the average channel. In the RELAP model, however, a hot pin is assumed in the average hydraulic channel, which has a higher power than the average rod, the hot pin factor being taken from Tab. 2.11.

Due to the various differences in the models and the details of the transient represented, only the first few seconds until the cladding temperature peak can be compared to some extent. Figure 4.24 shows the results for the cladding temperatures obtained with the three codes, the results obtained with APROS and RELAP showing the peak values of the hottest pin and the results with SMABRE showing the temperature of the average rod.

It is observed that the calculated hot pin temperature calculated by RELAP is much lower than the one calculated with APROS, as an effect of the negative reactivity coefficient. This peak temperature calculated by RELAP is lower than the average value calculated with SMABRE, although the same transient without scram was simulated. The differences in the results are related to a different time history of the thermal power.

In any case, the maximum peak cladding temperature was less than 1000°C. As a design limit for the total LOFW event without scram, one could assume that the acceptable limit would be 1200 °C, which is significantly more. Considering the different results obtained with the different codes, however, the questionable application of point kinetics and the uncertainties in the heat transfer coefficient, the simulations shown here cannot exclude the risk that this limit could eventually be exceeded. Thus, results shown so far are still considered to be rather preliminary.

4.3.4 Core coolability in case of loss of coolant

In case of loss of coolant (LOCA), special challenges stem from the transition from the supercritical single flow regime to subcritical two-phase flow conditions, and from the missing coolant recirculation in the core.

In fact, due to the strong density ratio between water and supercritical fluid, the core would empty very quickly in case of depressurisation and loss of mass through a break, and the once through coolant system could hinder or delay the refilling of the entire core.

In order to prove that the core could be cooled by means of safety injection systems under all circumstances, transients with breaks in both steam and feedwater lines have been analysed.

Small break LOCA with break in one of the main steam lines

The accident was assumed to be initiated by a 2.4% break in one of the main steam lines (classified as DBC4). This transient has been investigated using the SMABRE code.

An automatic depressurisation has not been activated in this case. As an additional constraint, the feedwater supply was assumed to fail simultaneously and water was assumed to be injected at high pressure after 25 seconds by means of a diesel-powered auxiliary feedwater pump, capable to deliver the same flow rate as one of the main feedwater pumps. Although this assumption is highly unlikely, this analysis gives an indication of what flow rate would be needed to avoid core overheating and would keep the fuel in a safe state.

Figure 4.25 shows the maximum cladding temperatures in the three regions of the core during the postulated accident. It can be observed that the reactor remains safe during the transient with an injection of 250 kg/s. Whether such an injection at high pressure is available or ADS has to be activated is a design issue.

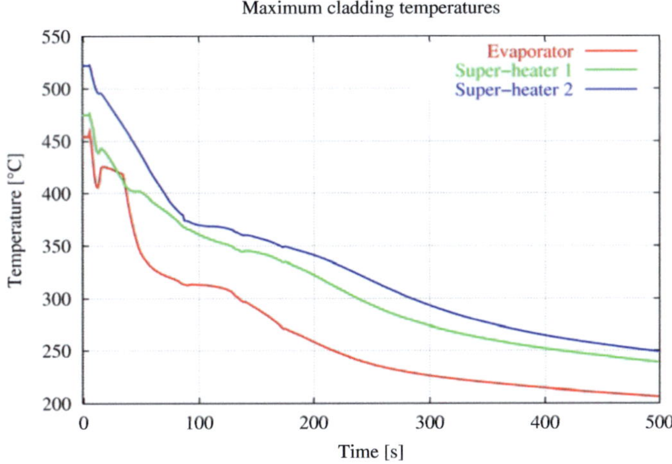

Figure 4.25: Small break in main steam line: maximum cladding temperatures in core sections [15]

Large break LOCA with break in one of the main steam lines

For this accident, it was postulated that a 100% break occurs in one of the main steam lines, just before the MSIV. This type of event is not expected to be very penalising for the HPLWR because the rupture of the steam pipe increases the flow rate inside the core and enhance the heat transfer between the fluid and the fuel cladding.

The accident is initiated by the full rupture of one steam line inside containments in 1 ms. LPCI is activated automatically when pressure drops below 6 MPa, and injects water from the pressure suppression pool at a temperature of 40°C and constant flow rate of 250 kg/s until the end of the transient.

This event has been simulated with two codes, with partly different boundary conditions and assumptions:

- CATHARE: the simulation was performed representing the hot channel (including the hot channel factors, Tab. 2.11) and considering a neutronic feedback. When steam flow reverses at one of the intact steam lines, the MSIV are closed in 1 s, the SCRAM signal occurs with a delay of 0.6 s and the duration of the control rod insertion is 3.5s. When the control rods are fully inserted, it is assumed that the overall reactivity of the core remains negative, even during the refilling phase.

- APROS: no neutronic feedback has been considered. The low pressure signal, sent 0.5 s after the pressure drops below 20.0 MPa, initiates the reactor SCRAM sequence and closure of the MSIV. The reactor power decreases according to a decay heat curve. The MSIVs close within 4 s from the break. Two cases have been considered, assuming that the feedwater pumps are running or not.

Just after the rupture of the steam line, the power starts immediately to decrease because of the density reactivity effect. The increase of the pressure gradient inside the core makes the flow rate increasing inside the core, which enhances the heat transfer between the fluid and the fuel cladding. The fuel cladding temperature starts to decrease immediately, and keeps on decreasing after the SCRAM signal. After LPCI activation, the simulations with the two codes show only minor differences. In the following, the results with CATHARE will be taken as reference, because CATHARE used a reactivity feedback for the core power and included both average and hot channels.

The essential results obtained with the CATHARE code are shown in Figures 4.26 and 4.27 [15].

The time history of the void fraction in superheater 1 (which is the last region to be refilled) shows that the entire core can be refilled within 1200 s with a LPCI injection rate of 250 kg/s. Long periods of low flow and flow reversal are calculated to occur in the hot or average channel and the cladding temperatures reflect these variations. In fact, small increases for long periods of time in the average channel and short duration peaks in the hot channel can be observed in the time history of the maximum cladding temperature in the evaporator and superheater 1. Figure 4.27 shows the temperatures in the average and hot channels of superheater 1. Figure 4.28 zooms into the first 10s, showing the peak cladding temperatures in superheater 1 and evaporator.

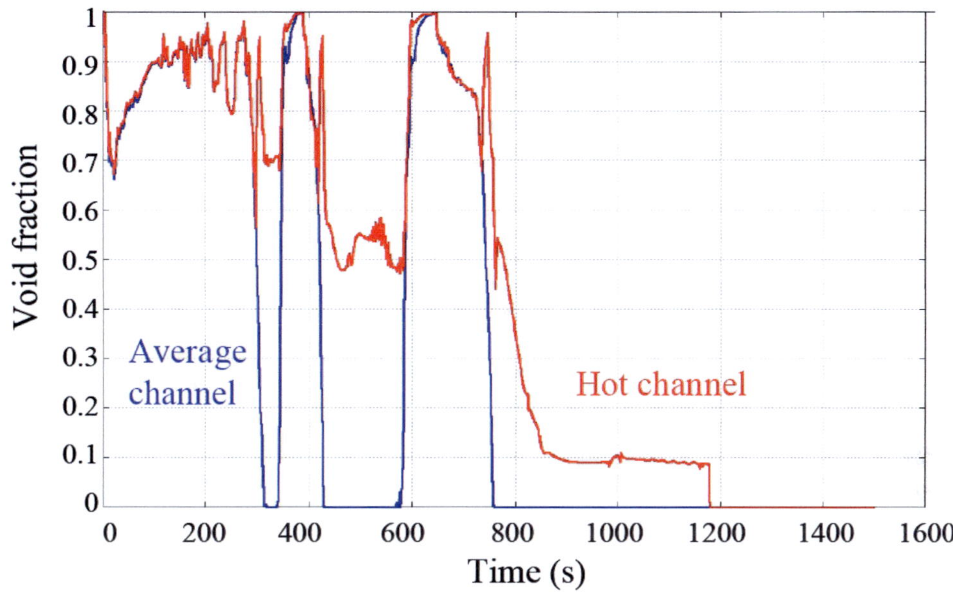

Figure 4.26: Average void fraction in superheater 1 calculated by CATHARE for a large break in the main steam line [14]

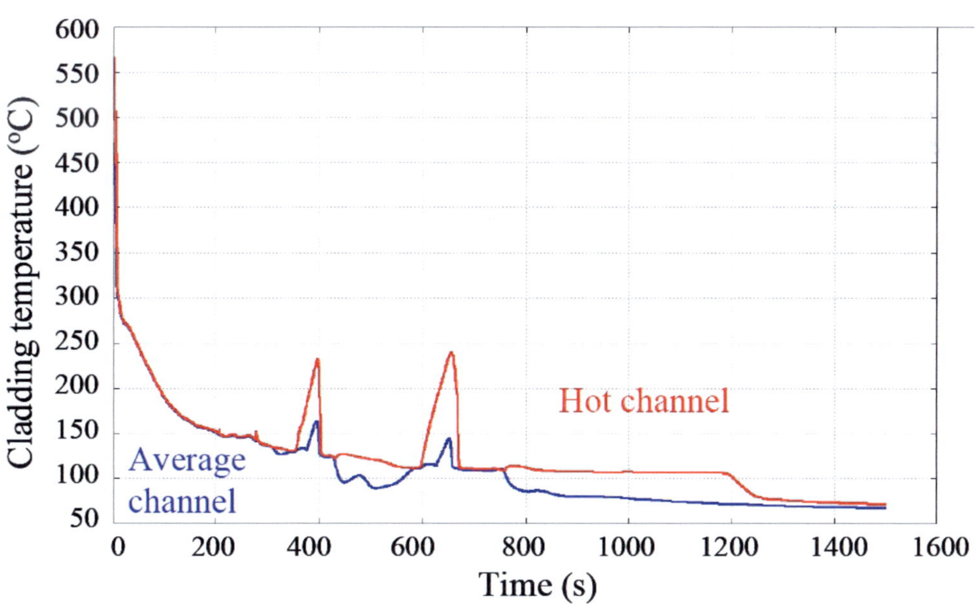

Figure 4.27: Maximum cladding temperature in superheater 1 calculated by CATHARE for a large break in the main steam line [14]

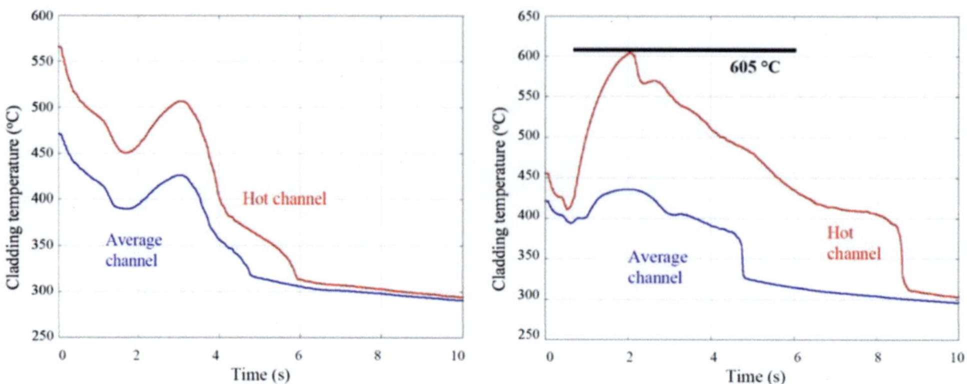

Fig. 4.28: Peak cladding temperatures in superheater 1 (left) and evaporator (right) during the first 10s after the break of the main steam line, predicted with CATHARE [14]

Qualitatively similar results have been obtained with the APROS code, with minor differences between the cases with and without continuing operation of the feedwater pumps. The only important difference is that for the case with pumps running a very high void fraction is still prevailing in the upper region of superheater 2 at the end of the calculation. A parametric study showed, however, that increasing the LPCI flow rate to 400 kg/s would lead to complete refilling of the core within 1000s. It could be therefore concluded that in case of large break in the main steam line, the LPCI is capable to maintain the core in a safe status. A flow rate of 250 kg/s can be expected to be sufficient to refill the entire core within a reasonably short time, and 400 kg/s provides a comfortable safety margin.

Large break LOCA with break in one of the feedwater lines

This type of event, classified as DBC4, had already been identified by Antoni and Dumaz [16] as the most penalising one as it induced backflow in the core and quasi-adiabatic core heat up after the automatic closure of the main steam isolation valves. They suggested implementing an automatic depressurisation system to restore the flow circulation inside the core and keep on cooling the fuel cladding during depressurisation. The depressurisation during this accident is very fast, with the system going from supercritical to subcritical range within a very short time. This event has been analyzed with the CATHARE code [15], considering the hot channel factors, Tab. 2.11.

The accident is initiated by the full rupture of one feed-water line inside containments in 1 ms. When steam flow reverses at one of the steam lines, the MSIV are closed in 1 s, the SCRAM signal occurs (the duration of the control rod insertion is 3.5s) and the sparger valves are opened with a delay of 0.6 s. LPCI is activated automatically when pressure drops below 6 MPa, and injects water from the pressure suppression pool at a temperature of 40°C and constant flow rate of 250 kg/s until the end of the transient.

Just after the rupture of the feedwater supply line, the power starts immediately to decrease because of the negative density reactivity effect. The inversion of the pressure gradient inside the core makes the flow rate going down, which deteriorates the heat transfer between the fluid and the fuel cladding. The period of increasing cladding temperature lasts a few seconds until the control rods are fully inserted after the scram signal and the spargers are fully opened. Then the flow rate inside the core becomes sufficient to cool again the core and the cladding temperatures remains very low whilst the depressurisation process is going on. After LPCI activation, when pressure gets very low, a flow reversal in the hot or average channel may occur and the cladding temperatures increase a little for a short period of time. At the end of the simulation, the core is refilled again.

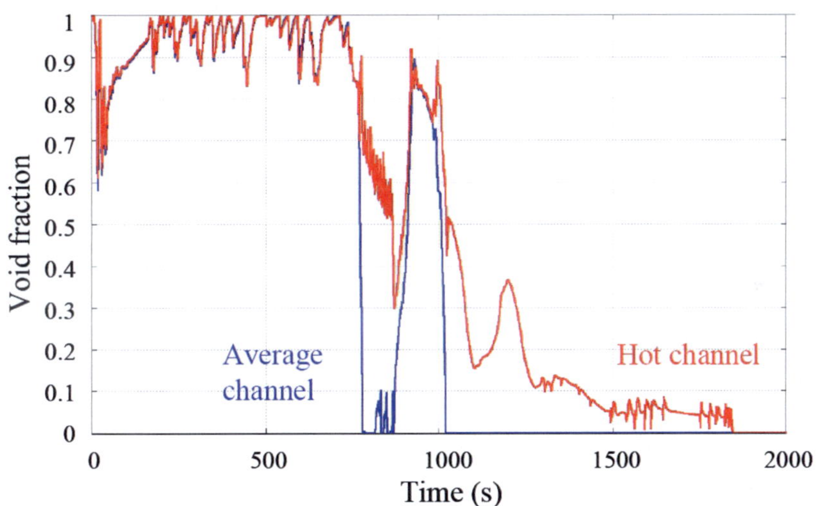

Figure 4.29: Average void fraction in superheater 1 calculated with CATHARE for a large break in the feedwater line [14]

Figures 4.29 shows that superheater 1 (the last region of the core to refill) is completely filled with water at the end of the calculation.

Figure 4.30 shows the time history of the maximum cladding temperature in superheater 1 for the hot and average channel. The peak temperature during the refilling phase is certainly acceptable.

These results have to be taken with some caution, because no model for counter-current flow limitation has been applied, due to the lack of specific information relevant for the complex geometry of the mixing chambers. However, in consideration of the flow reversals and the fact that superheater 1 is the region of the core that remains filled with steam for the longest time, especially a counter-current flow limitation in the upper mixing chamber above superheater 1 could hinder or delay the downwards flow of water and therefore the refilling of the core.

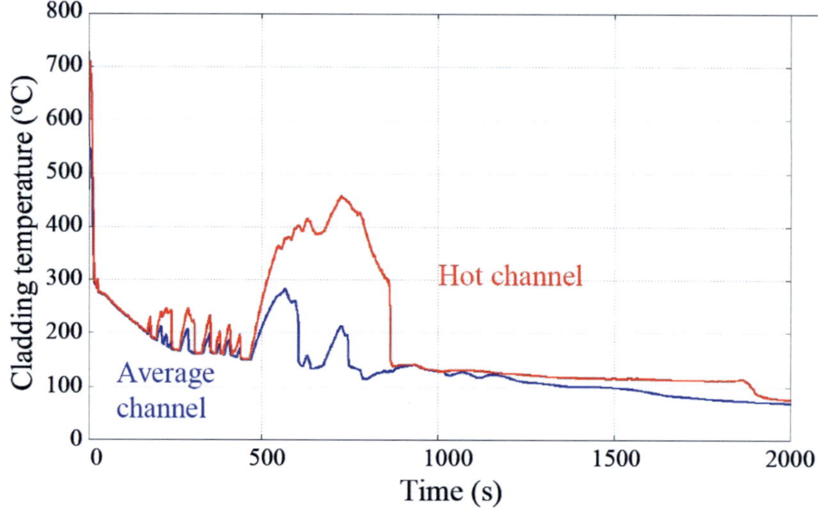

Figure 4.30: Peak cladding temperature in superheater 1 calculated by CATHARE for a large break in the feedwater line [14]

In conclusion, the results of the analyses show that the use of ADS system seems to be sufficient to limit the cladding temperature excursion to a reasonable value after the pressure has dropped below the critical value.

However, for this transient, the first few seconds have also to be looked at. In fact, the first temperature excursion occurs during the time when the pressure is still supercritical. In fact, Fig. 4.31 shows that peak cladding temperature in the hot channel reaches about 850 °C, which is far below the acceptable limit (~1200 °C) for such accidents.

These results have to be taken with some caution because of two reasons:

The calculation has been performed using the Jackson correlation [17] multiplied by a factor 0.9 for heat transfer, which is expected to provide a rather low value for the heat transfer coefficient in the normal heat transfer regime. However, in addition to the uncertainties in the heat transfer coefficient for the normal heat transfer regime, the eventual occurrence of heat transfer deterioration could cause the simulation to underestimate the initial temperature excursion. For clarification, Fig. 4.32 shows the ratio between the heat flux ϕ and the criterion for heat transfer deterioration ϕ_{HTD} by Yamagata et al. [18]

confirming that such deterioration of heat transfer must be expected at low flow rates and thus low mass flux G during the first seconds.

These numerical simulations of the transition from supercritical to subcritical pressure have never been assessed against experimental results, due to the lack of a database. It is therefore still uncertain whether the correlations used for steady state heat transfer are valid during the transition, or would substantially overestimate the cooling effectives of the developing two-phase mixture.

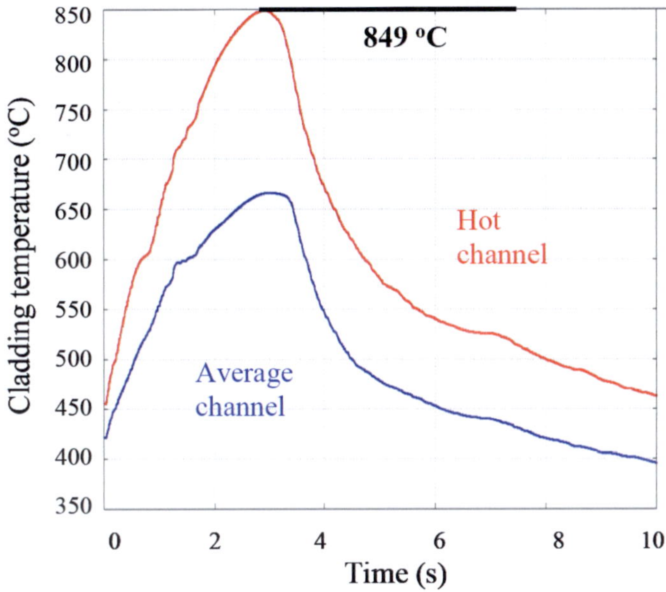

Figure 4.31: Peak cladding temperature in the evaporator calculated by CATHARE for a large break in the feedwater line during the first seconds [14]

In conclusion, the issue of an eventual overheating of claddings during the first seconds of an accident initiated by a large break in one of the feedwater lines is still open, and only experiments can clarify whether the model implemented in the codes are adequate for simulating transient heat transfer for supercritical conditions and during the transition from supercritical to subcritical region.

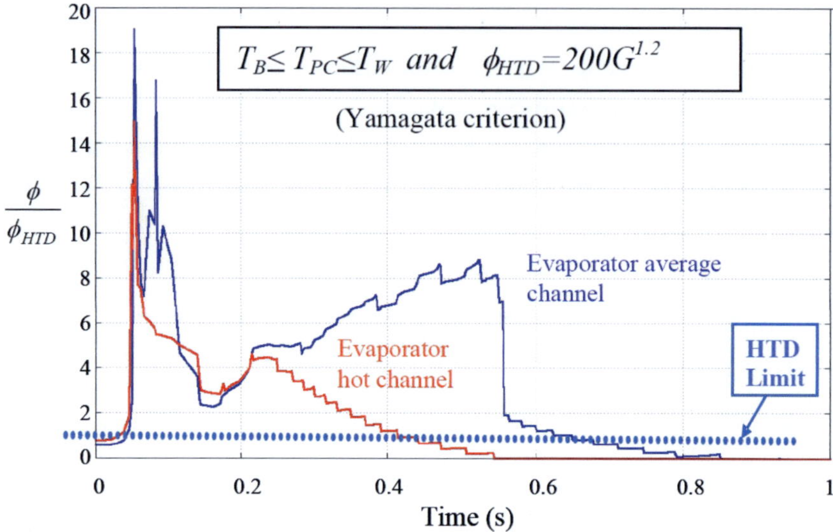

Figure 4.32: Occurrence of heat transfer deterioration (HTD) in the evaporator during the first second of a LOCA with break in the feedwater line [14]

4.3.5 Consequences of a control rod ejection

The accident is defined as the mechanical failure of a control rod mechanism housing such that the reactor pressure ejects a control rod up to its uppermost position. The consequence of this failure is a fast and large positive reactivity insertion which results in a core a power excursion with a large localized relative power increase. It is necessary to analyze the accident to determine if there is any fuel damage or damage to the reactor coolant system pressure boundary as a result of the power excursion.

This event has been analyzed with two coupled codes, ATHLET-KIKO3D [20] and SMABRE/TRAB-3D [15].

In the ATHLET-KIKO3D model, the primary circuit of the HPLWR was modeled by 88 thermo-fluid and 59 heat conduction objects. In the KIKO3D core model, the 156 clusters were individually represented, and each cluster consisting of 9 assemblies was divided into 20 axial nodes, both for neutronic and thermal-hydraulic analyses. The KIKO3D nodalisation of the three-pass core, as well as the position of the control rods are shown in Fig. 4.33. In this figure, the position of the ejected control rod and the clusters where the hottest temperatures were reached during the transient (defined as "hot channels", H-CH) are also indicated.

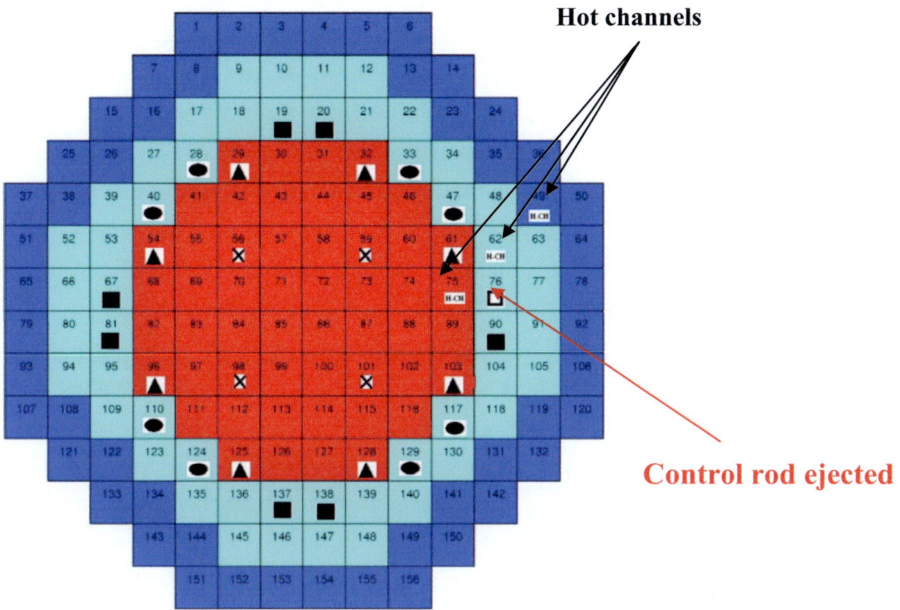

Figure 4.33: The KIKO3D nodalization of the three path reactor: 52 clusters in the Evaporator (Red), 52 clusters in the Super Heater 1(Light blue), 52 clusters in the Super Heater 2(Dark blue). Positions of the control rod groups and potential positions of the hot channels [15]

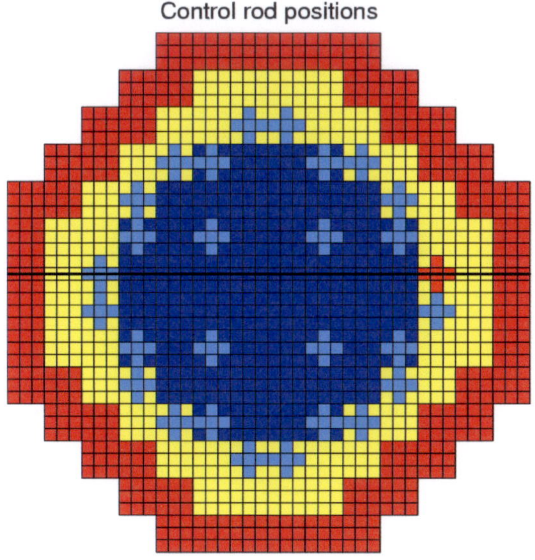

Figure 4.34: Core model for TRAB-3D and control rod positions. The ejected rod is shown in red and the position of axial cross-section for figure 4.4-6 is shown with a line

The input for the 3D core analysis code TRAB-3D includes the detailed description of the core with fuel rods and an individual flow channel attached to each fuel assembly. In the SMABRE input, the core is described with only several sectors and axial nodes. Three different core nodalizations were created for TRAB-3D for testing: using 3 hydraulic channels (1/core section), 156 hydraulic channels (1/cluster) and 1404 hydraulic channels (1/fuel assembly). The model with 1404 hydraulic channels proved to be the most stable and therefore it is used in all calculations presented here. Except for the core region, the same SMABRE model is used for reactor and steam cycle as in stand-alone SMABRE simulations. Figure 4.34 shows the control rod positions in blue, the ejected one being marked in red.

A sufficient number of initial reactor states to completely check all operational conditions must be analyzed to assure examination of the wide range consequences of the possible damage. In this analysis only two cases were investigated. The ejected control rod is situated in superheater 1 close to the evaporator which is an asymmetric position. The rod is ejected from 0 cm to 420 cm within one second in Case A and in 0.1 seconds in Case B.

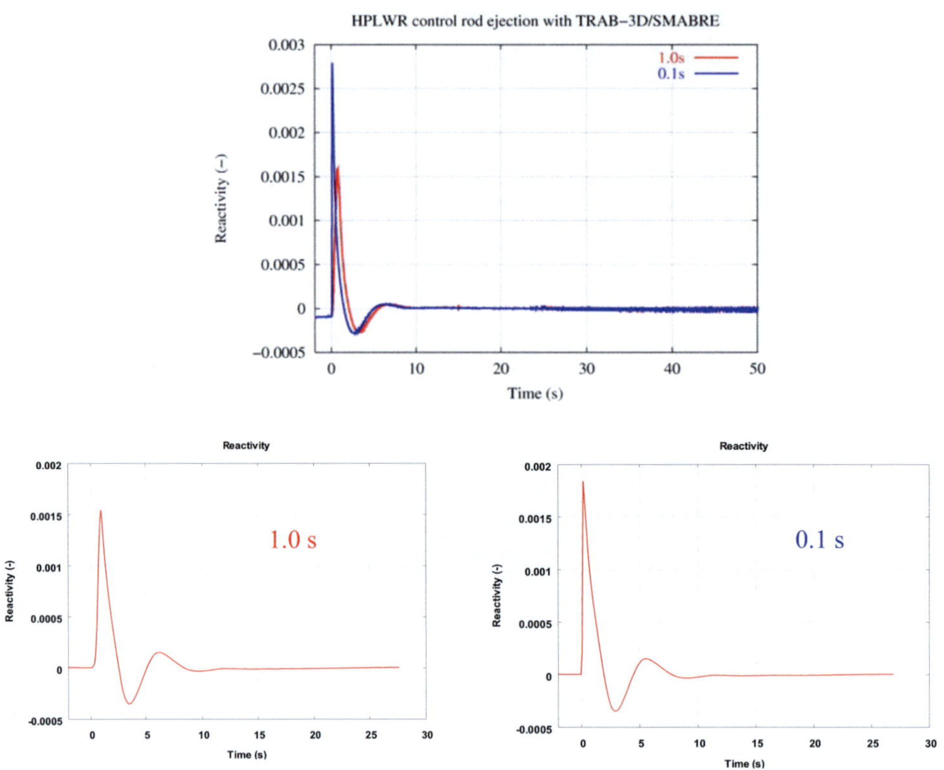

Figure 4.35: Comparison of the reactivity excursion calculated for a control rod ejection with SMABRE/TRAB-3D (top) and ATHLET-KIKO3D (bottom) [15]

Since the thermal-hydraulic model is very different in the two codes, fluid, cladding and fuel centreline temperatures are difficult to compare. Therefore, the comparison is here limited to the global variables that are directly affected by the 3-D neutronics, namely reactivity and power.

Figure 4.35 shows the comparison of the reactivities calculated by the two codes.

For the slow ejection case (1s) the results are practically identical. In the case of fast ejection (0.1 s) SMABRE/TRAB-3D predicts a somewhat larger insertion of reactivity, but the peak is of very short duration.

Consistent with the reactivity, the peak in the power calculated with SMABRE/TRAB-3D is higher than the one calculated with ATHLET-KIKO3D. In general, however, the results are in excellent agreement with each other and provide sufficient confidence that the results obtained for other transients are representative of the behavior of the HPLWR.

The sudden control rod movement causes a strong perturbation of the power distribution, which is later flattened by the feedback to great extent. The changes in the radial and axial power distributions calculated by TRAB-3D are shown in Figs. 4.36 and 4.37. The strong inhomogeneity of the local power excursion results in strong differences in cladding and fuel temperature increases in the various zones of the core. The results obtained with ATHLET-KIKO3D, where the hot channels are represented, permit to evaluate whether the acceptance criteria are fulfilled for this transient. Cladding and fuel centreline temperatures were examined in the hot channels of the three assemblies indicated in Fig. 4.33. Figures 4.38 and 4.39 show the maximum cladding temperatures in the three regions of the core. The maximum cladding temperature (860°C) exceeds slightly the criterion for buckling of the cladding but does not exceed the limit of 1200°C when severe oxidation of the cladding must be expected. The maximum centerline fuel temperature (2600°C) remains below the melting point of 2800°C, although it is not very far from it.

A closer look of the level of power achieved after the reactivity excursion (after 10 s) in the case of 0.1s ejection time shows that the difference between the two states (before and after control rod ejection) is 15 to 20 % larger in the calculation with SMABRE/TRAB-3D.

Although this difference in the total power does not reflect necessarily the changes in the hottest part of the core (the Evaporator), this result raises the question whether the fuel temperature acceptance limit could be exceeded. In fact, in the calculation with ATHLET-KIKO3D, the fuel centreline temperature after 25 s (Fig. 4.39) was very close to the limit, and an uncertainty in the power of more than 10% would result in an uncertainty in the pellet temperature that practically would eliminate any safety margin for the case of a control rod ejection transient.

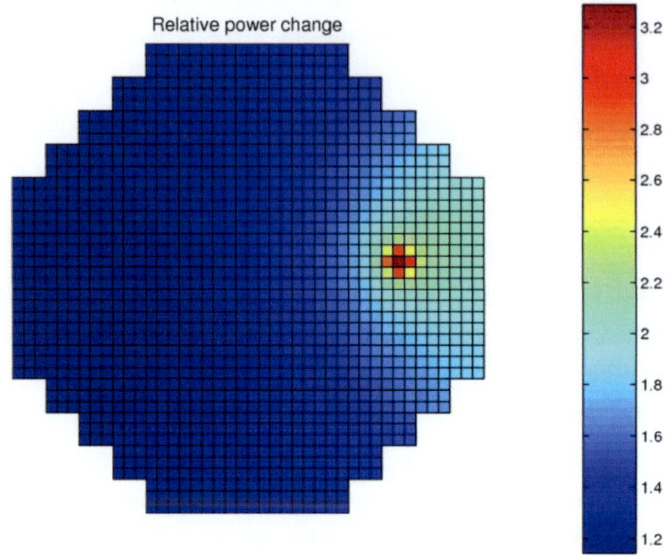

Figure 4.36: Relative power change from initial power to maximum power during control rod ejection in 1.0s calculated with TRAB-3D [15]

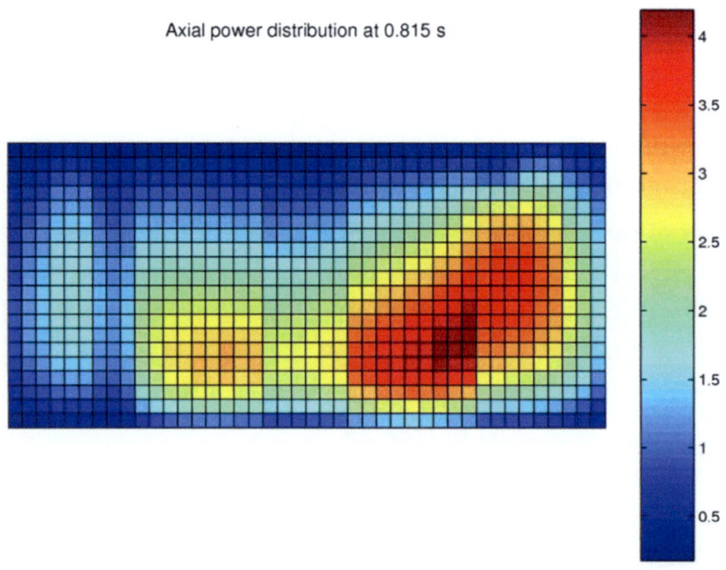

Figure 4.37: Axial power distribution at the time of maximum power in the control rod ejection transient in 1.0s calculated with TRAB-3D [15]

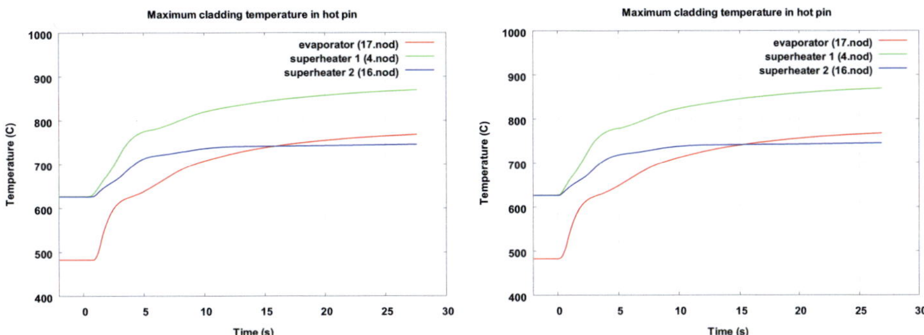

Figure 4.38: Maximal cladding temperature and its position in the hot channel of the evaporator, the superheater 1 and 2 during the transient (Case A: ΔT=1 s / Case B: ΔT=0.1 s), predicted with ATHLET/KIKO3D [15]

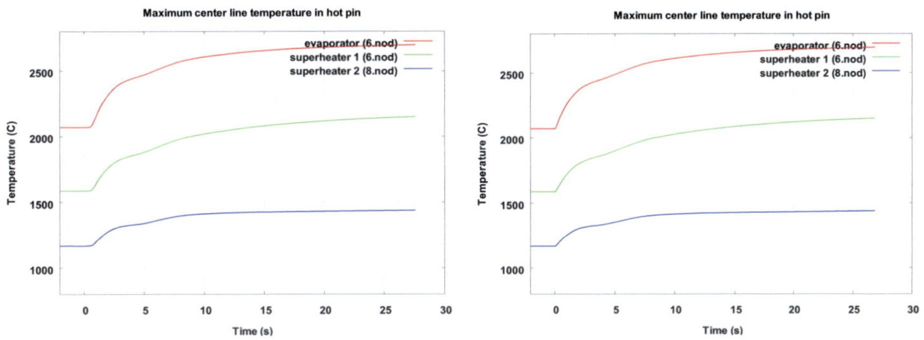

Figure 4.39: Maximal centerline temperature and its position in the hot channel of the evaporator, the superheater 1 and 2 during the transient (Case A: ΔT=1 s / Case B: ΔT=0.1 s), predicted with ATHLET/KIKO3D [15]

4.3.6 Other events

Analyses of other events investigated with ATHLET-KIKO3D are summarized in Table 4.6. The events are classified as events with anticipated operational occurance (DBC2) and postulated accidents (DBC3/4) for which different limits for cladding surface temperature and fuel centerline temperature were defined in Tab. 4.1.

The acceptance criteria given in Tab. 4.1 are fulfilled in all but one case. However, in spite of the relatively strong feedback, the hot channel temperatures are not far from the limits in some cases, which points out the necessity of detailed transient analyses. The reasons are the strong sensitivity of the neutronic characteristics on the moderator density, changing in a wide range and, as a consequence, the significant perturbation of power peaking factors during the transients.

In case of uncontrolled withdrawal of an absorber from the bottom position without scram, fuel melting is predicted which can be avoided by limiting the allowed position of the control rods or by applying lower worth regulating rods compared to shutdown rods. Some further study is needed to introduce appropriate preventing measures.

Initiating event	Classification of the initiating event	Maximum cladding surface temperature	Maximum pellet centerline temperature
Uncontrolled absorber group withdrawal from the bottom position with scram	DBC2	705 °C	2250 °C
Uncontrolled absorber group withdrawal from the middle position without scram	DBC2	730 °C	2150 °C
Uncontrolled withdrawal of an absorber in asymmetric position without scram	DBC2	1000 °C	3100 °C
Control rod ejection	DBC4	860 °C	2600 °C
Loss of feedwater heating	DBC2	690 °C	2400 °C
Control rod malfunction: (case A and B)	DBC2	648 °C	2050 °C

Table 4.6: Maximum cladding and fuel centreline temperatures for the various events calculated with ATHLET-KIKO3D [15]

4.3.7 Consequences for safety system design

The methodology used for the safety analyses was based on the decoupling from the still unresolved design issue to reduce the initial hot channel factors and therefore the maximum cladding and fuel temperature under normal operation condition. This implies that the analyses presented in the previous chapters must be regarded as preliminary, and the results only show the response of the system, but give limited information on the fulfillment of acceptance criteria.

Bearing this in mind, one can derive from the analyses the following recommendation for the design of the safety systems and the core:

- The ADS system proposed on the base of the design calculations performed with a coarse model of the core is adequate to limit pressure excursions far below design limits. It has also been shown that the system with the chosen parameters enables the water introduced in the vessel to provide sufficient flow in the core and effective cooling of the fuel for all anticipated transients and accidents investigated. Therefore the set of parameters chosen for the safety relief valves (Flow area=0.09 m^2; actuation pressure=26 MPA; opening time=0.2 s) can be recommended for future safety analyses for an optimized core.

- In case of loss of feedwater by failure of one or two pumps, the temperature excursion is lower than 60 K, even when a very small inertia of the pump is assumed. Considering this small excursion, the initial temperature of the cladding (and therefore the core design) has a stronger impact on the fulfilling the acceptance criteria than the temperature excursion during this type of transients. Therefore, the pump inertia is not a critical parameter. It should also be taken into consideration that in case of a total loss of feedwater, and consequent pressure rise, the main feedwater isolation valves would close quickly, and the inertia of the pump would not play any role.

- For low values of the core flow rate (below 40% of the nominal value), flow reversal could occur in the gap and moderator channels. Modifications in the moderator flow path were needed to avoid this condition. Moreover, it has been observed that the heat transfer from the fuel to the moderator channels boxes plays an important role on the transient moderator flow pattern, and a better insulating material for the moderator boxes would be beneficial.

- The core coolability by means of the LPCI has been demonstrated for large break LOCAs in both the steam and feedwater lines. A mass flow of 250 kg/s is sufficient to avoid cladding overheating and most probably to ensure a fast refilling of the entire core. A larger value (400 kg/s) would provide additional safety margin with respect to the complete refilling of the core. The shortcoming of this system is the high power required to operate the pumps.

- In order to reduce the power demand for emergency cooling water injection systems, an active HPCI system has also been investigated, as discussed in chapter 4.1, Figure 4.7. Design calculations showed that this system should be capable to provide core cooling in the case of an event with depressurization. As it cannot be operated for a long time, an option would be to start the LPCI system at low pressure and close the HPCI system. This strategy, however, needs to be further investigated in future safety analyses.

- A passive safety system with a steam injector, as proposed in chapter 4.2, has not been analyzed yet with transient system codes. The question if this system will perform as expected is still open.

- Preliminary analyses indicate that in case of large break LOCA in the feedwater line, the backflow limiter was not useful or even had a detrimental effect on the initial core cooling because it reduced the core flow. Further studies will be necessary to decide whether this devise should be included or not in the final design.

- The limited analysis for a small break LOCA showed that the intervention of an auxiliary feedwater supply system (starting at nearly full pressure) with 250 kg/s injection rate would be sufficient to maintain the reactor in a safe state. Whether this system will be available or ADS should be actuated to permit the intervention of other systems will be a design choice.

- In case of accidents initiated by reactivity insertions, fuel centerline temperatures arrive at values very close to acceptability limits. The core design has to be optimized to avoid this condition.

- In one case, namely the uncontrolled withdrawal of an absorber from the bottom position without scam, fuel melting is calculated to occur. This can be avoided by limiting the allowed position of control rods or by applying lower worth regulating rods compared to shutdown rods. Some further study is needed to introduce appropriate preventing measures.

Two issues have been identified which are still pending and must be solved before being able to decide whether design modifications are needed:

- In case of a total loss of feedwater, especially if combined with the failure to scram the reactor (ATWS), the initial cladding temperature excursion could not be evaluated with confidence using point kinetics. Three-dimensional coupled codes are not available yet. Therefore, one of the first tasks of the safety analysis in a future project must be the analysis of this transient using appropriate analysis tools.

- Also in case of a large break LOCA in the feedwater line, the first seconds are critical. The results obtained with the system codes seem to indicate that the cladding temperature excursion is still acceptable. However, due to the lack of adequate correlations for deteriorated heat transfer and of assessment of the physical models for the conditions prevailing during the transition from supercritical to subcritical region, the peak cladding temperature is still very uncertain. Eventually, experimental data will be needed to qualify the codes for these conditions and consequently provide confidence in their predictions.

4.4 Mechanical analyses of the containment

These analyses demonstrate that the containment becomes the most important barrier against release of radioactive material in case of an accident. It confines the primary system as soon as the containment isolation valves are closed. The HPLWR containment has been designed to stand an internal pressure increase up to 0.5 MPa, a horizontal ground acceleration of 0.8 m/s^2 in case of an earthquake and an airplane crash or an outside explosion pressure wave. The latter external events have been decoupled from the containment by a gap between the reactor building and the containment, such that the reactor building will take over the mechanical load but the containment inside will be unaffected. The mechanical structure of the containment has been analyzed by Maisch and Siegel [21].

Under normal operating conditions, the containment shown in Fig. 4.8 has to carry the following loads, as sketched in Fig. 4.40:

- The weight of the pressure suppression pool with a liquid height of 4.85m.

- The weight of the 4 upper pools with a liquid height up to 8.80m in case of depressurization.

- The weight of the reactor acting via the support structure of the pressure vessel on a console of the containment structure, which is 1510t if reactor and containment are closed and around 2100t if the reactor is opened and the drywell above the core is flooded.

- The weight of the pool above the containment with a height of up to 13m.

- The weight of the concrete of the containment itself and of components of the safety system.

The pump room underneath the pressure suppression pool is separated by walls into 4 closed rooms, containing the LPCI pumps and the residual heat removal systems, to minimize the risk of a common failure of all these 4 systems. The walls are carrying the load of the pressure suppression pool. Four penetrations from the outside to these rooms are foreseen to get access to each room individually.

The weight of the pressure vessel is carried by the console, marked in green in Fig. 4.41, left, and the bending moment of the 4 upper pools is taken by a pressure ring connecting the walls of the upper pools, marked in green in Fig. 4.41, right.

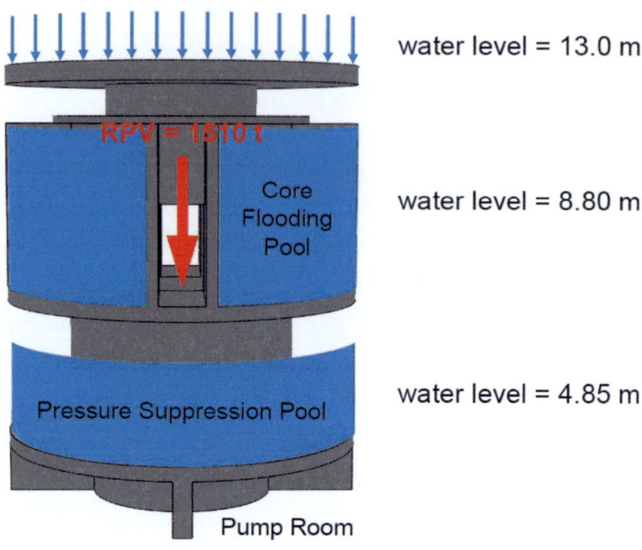

Fig. 4.40: Mechanical loads acting on the containment structure

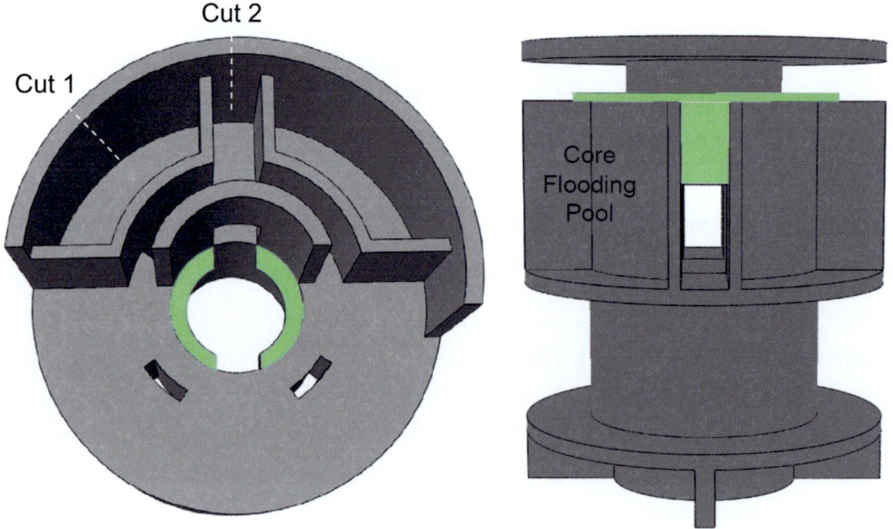

Fig. 4.41: Support console for the reactor pressure vessel, marked in green (left) and pressure ring of the upper pool, marked in green (right) [21]

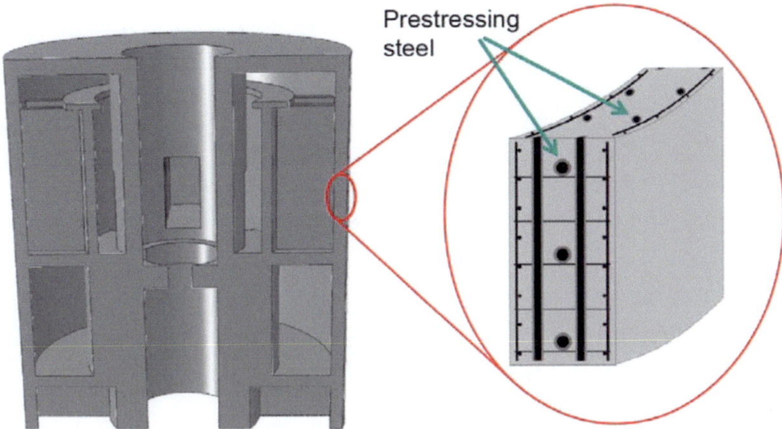

Fig. 4.42: Reinforced, pre-stressed concrete structure of the containment [21]

A gate with a lock from outside to the drywell is foreseen between two upper pools, indicated in Fig. 4.8 on the left hand side underneath the containment isolation valves. From there, access to the pressure suppression pool is foreseen from the top.

The outer containment structure is designed with reinforced, pre-stressed concrete as sketched in Fig. 4.42.

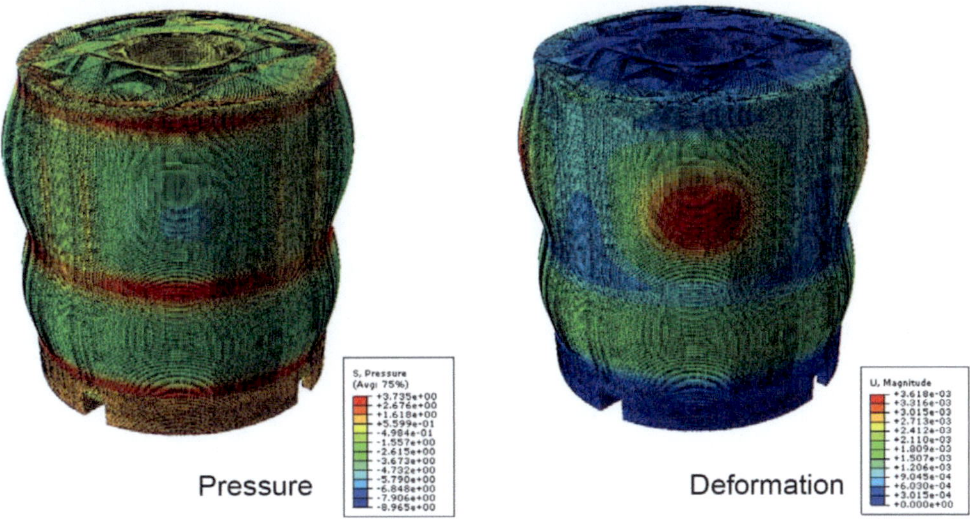

Fig. 4.43: Stresses [MPa] and deformations [m] of the containment under an internal pressure of 0.5 MPa [21]

Static and dynamic stresses and deformations have been analyzed with the finite element code ABAQUS. Some preliminary results are given in Figs. 4.43 to 4.45 to illustrate the expected response to the applied loads.

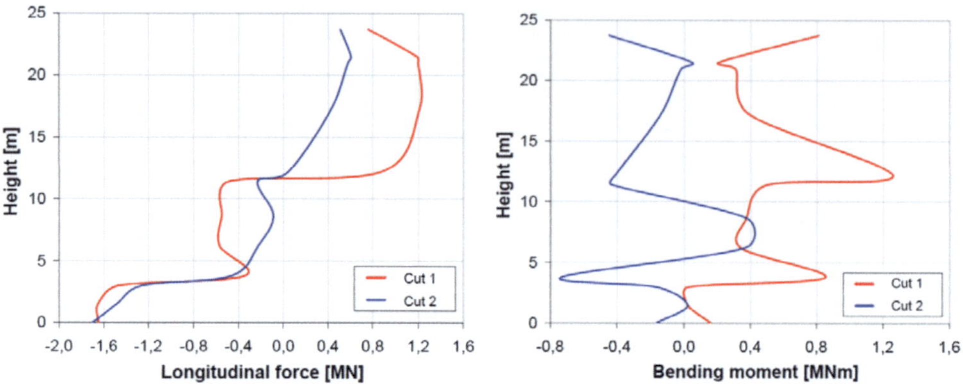

Fig. 4.44: Longitudinal force and bending moment of the cylindrical pre-stressed concrete structure [21]. Position of cuts indicated in Fig. 4.41.

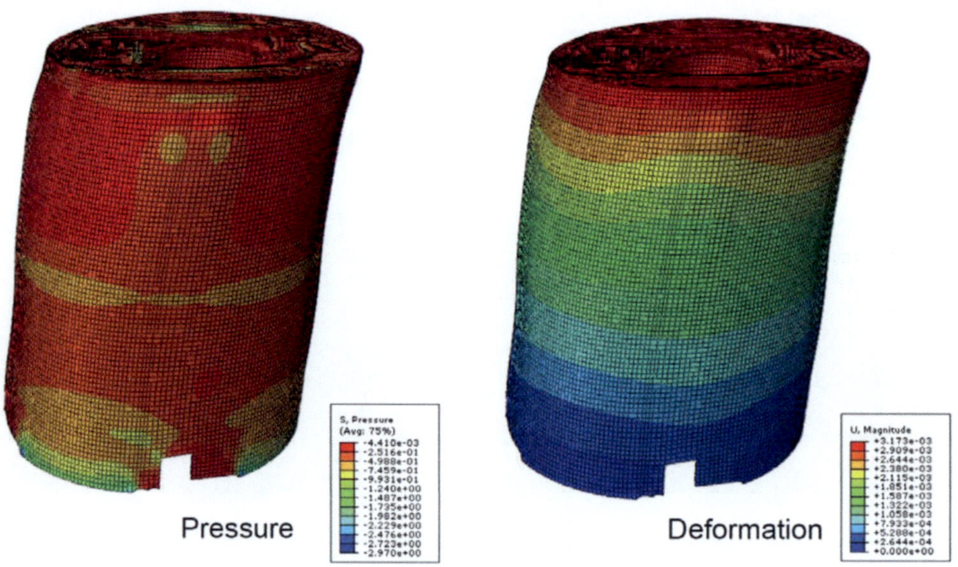

Fig. 4.45: Stress [MPa] and deformation [m] amplitude of the containment due to an earthquake with a ground acceleration of 0.8 m/s^2 [21]

Fig. 4.43 shows stresses and deformations of the containment under an internal pressure of 0.5 MPa. The solid floors are keeping the containment diameter at the initial size, while the cylindrical structure in between is deforming up to 3.6 mm to the outside. A metal liner will be required to keep the containment leak tight. The longitudinal force and the bending moment of the outer cylindrical structure are plotted in Fig. 4.44 as a function of height

The stress and deformation amplitude caused by an earthquake with a ground acceleration of 0.8m/s^2 is shown exemplarily in Fig. 4.45, reaching up to 3.2mm displacement at the top of the containment.

All these preliminary results are based on the simplified assumption of a homogenous concrete material with typical average properties. More detailed analyses will be required once the containment design has been worked out in more detail.

4.5   Discussion

Similar as the design of the core and of the primary system, the safety system design cannot be considered to be final either and results achieved up to now should rather been seen as exemplarily. So far, analyses were primarily concentrated on the proposed active safety systems, and the design of passive safety systems is still rather a vision than a concept. Moreover, none of the system codes which were used could describe all physical phenomena simultaneously. While only two codes could manage the depressurization to sub-critical pressure, other codes were superior in modeling the dynamics of the local neutron flux and the reactivity feedback of the local coolant density. And finally, none of the codes can be considered to be validated for this application.

Nevertheless, the work done so far is an important step towards the development of supercritical water cooled reactors. As this reactor type does not include a closed cooling loop in the primary system any more, the safety strategy had to be reconsidered in general. As suggested by Oka et al. [8], the coolant inventory in the primary system is not a safety relevant parameter any more, but rather the coolant flow rate and the coolant inventory in the containment. As the core can only be cooled sufficiently after containment isolation once the interrupted cooling loop is closed again, a depressurization of the reactor into the containment and reinjection of the missing coolant from the containment pool turned out to be an effective strategy to avoid excessive temperature peaks in the core. The various transient analyses performed up to now confirm that major damages of the core can indeed be avoided under accidental conditions. Even in case of a break of the feedwater line, an additional coolant discharge into to containment is meaningful to maintain a high coolant flow rate through the core, despite the fact that even more coolant is lost then. The peak temperatures which were predicted indicate that the active safety systems will perform similar as with reactors of the 3rd generation. There is no indication, however, that the HPWLR will be safer than 3rd generation light water reactors.

A significant advantage of the HPWLR containment is rather to be expected for its costs. For illustration, Fig. 4.46 shows a size comparison of the containment of the AP1000, the boiling water reactor Gundremmingen and the HPLWR. With only 21.5m outer diameter and 25m outer height, the HPLWR containment is significantly smaller, but still containing more than $2000m^3$ of water. The main reason is the missing steam generators or missing steam separator, like with PWR or BWR respectively, so that the containment size could be reduced to the minimum which is just required to fulfill the safety requirements.

The development of passive safety systems, on the other hand, will be more challenging. In particular, the HPLWR core with its upward and downward flow path is not well suited for a natural convection loop through a containment condenser, so that other options should be considered instead. The proposed steam injector is just one of several ideas which should be studied in more detail in future projects.

Moreover, the severe accident management strategy has not been worked out in detail either. Progression and retention of a core melt will need to be studied at a later stage of development.

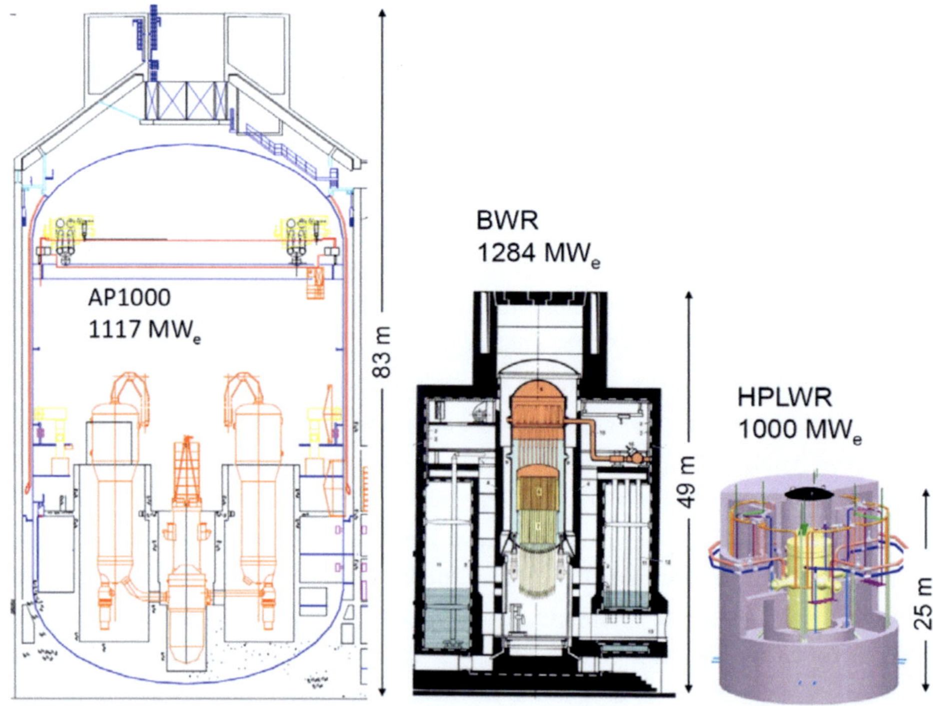

Fig. 4.46: Size comparison of the AP1000, the BWR Gundremmingen and the HPLWR containment

4.6 References

[1] A Technology Roadmap for Generation IV Nuclear Energy systems, issued by the US DOE Nuclear Energy Research Advisory Committee and the Generation IV International Forum, GIF-002-00, December 2002

[2] European Utility Requirements for LWR Nuclear Power Plants, Volume 2: Generic nuclear island requirements, Chapter 2.1: Safety Requirements, Revision C, April 2001

[3] J. Hamel, EPRI Utility Requirements Document for next generation nuclear plants, Revision 10

[4] D. Bittermann, T. Schulenberg, M. Andreani, The safety concept of the HPLWR, 4th Int. Symp. on SCWR, Paper 11, Heidelberg, Germany, March 8-11, 2009

[5] T. Schulenberg, D. Visser, Thermal-hydraulics and safety concepts of supercritical water cooled reactors, Proc. NURETH 14, Paper 635, Toronto, Canada, Sept. 25-29, 2011

[6] Sempel AG, www.sempell.com

[7] M. Schlagenhaufer, T. Schulenberg, J. Starflinger, D. Bittermann, M. Andreani, Design proposal and parametric study of the HPLWR safety system, Proc. ICAPP 10, Paper 10152, San Diego, USA, June 13-17, 2010

[8] Y. Oka, S. Koshizuka, Y. Ishiwatari, A. Yamaji, Super light water reactors and super fast reactors, Springer 2010

[9] Y. Ishiwatari, C. Peng, T. Sawada, S. Ikejiri, Y. Oka, Design and improvement of plant control system for a super fast reactor, Proc. ICAPP 09, Paper 9261, Tokyo, Japan, May 10-14, 2009

[10] B. de Marsac, D. Bittermann, J. Starflinger, T. Schulenberg, Containment design proposal with active and passive systems for an High Performance Light Water Reactor, 4th Int. Symp. SCWR, paper 10, Heidelberg, Germany, March 8 – 11, 2009

[11] P. Dumaz, G. Geffraye, V. Kalitvianski, E. Verloo, M. Valisi, P. Méloni, A. Achilli, R. Schilling, M. Malacka, M. Trela, The DEEPSSI project, design, testing and modeling of steam injectors, Nucl. Engineering and Design 235, p. 233-251, 2005

[12] B. de Marsac, Heat removal system and containment integration for the High Performance Light Water Reactor, Studienarbeit, Institute for Nuclear and Energy Technologies, Forschungszentrum Karlsruhe, 2008

[13] M. Schlagenhaufer, Simulation des Dampf-Wasserkreislaufs und der Sicherheitssysteme eines High Performance Light Water Reactors, Dissertation Karlsruhe Institute of Technologies, 2010

[14] M. Andreani, D. Bittermann, Ph. Marsault, O. Antoni, A. Keresztúri, M. Schlagenhaufer, A. Manera, M. Seppäla, J. Kurki, Evaluation of a preliminary safety concept for the HPLWR, Progress in Nuclear Energy, 55, 68-77, 2012

[15] J. Starflinger, T. Schulenberg, P. Marsault, D. Bittermann, E. Laurien, Cs. Maráczy, H. Anglart, J.A. Lycklama, M. Andreani, M. Ruzickova, T: Vanttola, A. Kiss, M. Rohde, R. Novotny, High Performance Light Water Reactor Phase 2, Final Public Report, Assessment of the HPLWR Concept, Dec. 2010, www.hplwr.eu

[16] O. Antoni, P. Dumaz, Preliminary calculations of a supercritical light water reactor concept using the CATHARE code, Proc. ICAPP 03, Paper 3146, Cordoba, Spain, May 4-7, 2003

[17] J.D. Jackson, Consideration of Heat Transfer Properties of Supercritical Pressure Water in Connection with the Cooling of Advanced Nuclear Reactors", Proc. 13th Pacific Basin Nuclear Conf., Shenzhen City, China, Oct. 21-25, 2002

[18] K. Yamagata, K. Nishikawa, S. Hasegawa, T. Fujii, S. Yoshida, Forced Convection Heat Transfer to Supercritical Water Flowing in Tubes, Int. J. Heat Mass Trans., Vol. 15, 2575-2593, 1972

[19] M. Hänninen, J. Kurki, Simulation of flows at supercritical pressure with a two-fluid code, Proc. NUTHOS-7, Seoul, Korea, Oct. 5-9, 2008

[20] Cs. Maráczy, A. Keresztúri, I. Trosztel, Gy. Hegyi, Safety analysis of reactivity initiated accidents in a HPLWR reactor by the coupled ATHLET-KIKO3D code, Progress in Nuclear Energy 52, 2009, pp. 190–196.

[21] M. Maisch , S. Siegel, Institute of Reinforced Concrete and Building Materials, Karlsruhe Institute of Technology, unpublished, 2010

5 Steam Cycle and Layout of the Power Plant

Similar as with the primary system and the containment design, it has been the primary design target of the HPLWR to minimize the specific plant erection costs, i.e. the total costs divided by the net electric power. This is equivalent with maximizing the turbine power at a given plant investment. The HPLWR net electric power has been defined arbitrarily as 1000 MW, keeping in mind that a larger plant size will most likely decrease further the specific plant erection costs by scaling effects. As the costs of enriched uranium are still cheap, the net efficiency of the plant is only of secondary importance. Earlier steam cycle analyses of Oka et al. [1] indicated that a net efficiency of around 44% should be achievable at a live steam temperature of 500°C. The improvement of plant efficiency, however, is not proportional to the savings of fuel costs. We have to keep in mind that higher fuel enrichment and lower fuel burn-up will compensate the savings of fuel costs to some extend compared with a conventional BWR.

The main driver for lower specific plant costs is the higher steam enthalpy at the turbine inlet which reduces the required steam and feedwater mass flow rates needed for the given turbine power, which reduces the size of the medium and low pressure turbines, the condensers and the feedwater system in comparison with a BWR. With 500°C live steam temperature and 25 MPa live steam pressure of the HPLWR, the enthalpy difference between the reactor outlet and the condenser inlet with 5 kPa pressure and a steam quality of 89% is 1.8 times larger than the one of a BWR with 7 MPa saturated steam and same condenser conditions. Therefore, the feedwater mass flow rate per MW net power of the BWR (Gundremmingen) is larger by 32% than the specific mass flow rate of the HPLWR.

The use of activated steam in the once through steam cycle is a general drawback of the plant design, which is common with BWRs, or even worse as fission products will not be restrained in the liquid phase inside the reactor. This implies that the turbine building is another control area of the HPLWR power plant. Same as with BWR, however, the main source of activation during normal operation arises from decay of N-16 due to activation of O-16 in the core. The wetwell of the condenser should be designed sufficiently large to prevent activation of the feedwater system and thus to minimize the shielding effort there.

The HPLWR plant design was intended to serve as a load following capacity in the load range from 50% to 100% of the design power. Steady state operation at lower load is not expected to be economical, and the HPLWR design target is excluding this load range accordingly.

5.1 Design strategy

The basic strategy of the HPLWR plant concept has been to use the long term experience of pressurized water reactors and of boiling water reactors on one hand, and the experience with fossil fired power plants designed for supercritical steam conditions on the other hand, to derive a novel plant concept with a minimum of research needs. Therefore, we should have a closer look first on the design of latest coal fired power plants with supercritical steam conditions.

One of the latest and largest coal fired power plants in the world is unit K of the German power plant Niederaussem [2]. It is fired with lignite to produce a gross power output of 1012 MW and a net output of 950 MW with a net efficiency of 43.2%. With a life steam pressure of 26 MPa, a reheat pressure of 5.5 MPa and life steam and reheat temperatures of 580°C and 600°C, respectively, the gross power and pressures are comparable with the HPLWR. The power plant is in commercial operation since 2003.

The evaporator, which is the tube wall of the boiler, wherein coolant is passing the pseudo-critical point at supercritical conditions, features 2 intermediate mixing stages to homogenize the coolant while being heated up. The maximum heat flux, provided by the flames, is less than 400 kW/m^2 to limit the post-dryout wall temperatures below ~550°C. All boiler materials are high Cr ferritic-martensitic steels [3]. The evaporator is followed by 6 parallel separators which serve as a start-up system at sub-critical pressure such that a minimum mass flow rate can be maintained in the evaporator at low power. In case that liquid is separated there, it is collected in a vertical vessel and re-supplied to the feedwater line, upstream the economizer, via a recirculation pump. Four superheater stages with intermediate coolant mixing each and coolant injectors before the last superheater stage, minimizing the peak coolant temperature, bring the steam temperature up to 580°C. The reheater is designed as tube bundles inside the boiler house with 2 heat-up stages and a mixing and coolant injection system in between. A single flow high pressure turbine, a double flow medium pressure turbine and two double flow low pressure turbines form the turbine train running at full speed of 50Hz. The steam cycle is operated with sliding pressure above ~40% thermal power, such that the steam valves are kept open and the life steam pressure is decreasing almost proportionally with the feedwater mass flow rate while lowering the power. The high pressure steam mass flow rate at full power is ~800 kg/s. 10 preheater stages produce a feedwater temperature of 293°C at 31.2 MPa. A feedwater tank at 1.5 MPa serves as preheater 7.

Taking such a power plant design as a starting point for HPLWR development, we have to consider the following basic differences of the nuclear system:

Intermediate coolant mixing in the core like in the boiler has been tried to some extend with the 3 pass core concept as discussed in Chapter 3. Even such mixing, however, will not be as homogeneous as in the boiler design, so that hotter peak coolant temperatures must be expected in the reactor.

The peak heat flux of a fuel rod with 39 kW/m linear power and with an outer cladding diameter of 8 mm is 1550 kW/m^2, which is almost 4 times the maximum heat flux of the boiler tubes, so that a sliding pressure operation with post-dryout conditions in the core must definitely be avoided. As discussed by Behnke et al. [4], the problem associated with post-dryout conditions in the core is not only the hot cladding temperature, which might be acceptable at least in the very low load range, but also the temperature difference along the circumference of a fuel rod in case that dryout occurred on one side, but the cladding is still wetted on the opposite site. Such conditions will cause bending of the fuel rods, followed by coolant blockage on the hotter side, leading to further bending and finally to contact of the fuel rods, which is causing a burn-out of the fuel rods. Behnke [5] shows with a sub-channel analysis that a uniform dryout location will be rather unlikely.

Steam extraction to a start-up system at the evaporator outlet and a nuclear reheater inside the core are complicating the core design significantly, at least in a pressure vessel type reactor, so that other options must be considered for start-up and re-heat in the nuclear system.

The lower core outlet temperature of 500°C of the HPLWR design and limitation of the reheater (as discussed below) are easing the material requirements of the turbines, but result in a higher steam mass flow rate of ~970 kg/s at comparable turbine power.

Last but not least, the containment as the outer safety barrier, the protection of the activated steam in the entire steam cycle and a residual heat removal system are requirements which are not foreseen in any fossil power plant, causing some further basic differences.

Nevertheless, however, the basic technologies needed to build the nuclear supercritical steam cycle components are still considered to be similar as those of the fossil fired plant and synergies could be gained.

5.2   Thermodynamic options

The steam cycle of the HPLWR has been optimized by Brandauer et al. [6] for steady state, full load conditions using the thermodynamic code IPSE-Pro 4.0. Their result is shown in Fig. 5.1, which will be used as a reference here. Aiming at a net power output of 1000 MW$_e$, which was rather an arbitrary target definition than a result of an optimization, the reactor has to produce a thermal power of 2300 MW$_{th}$ which results in a net efficiency of 43.5%.

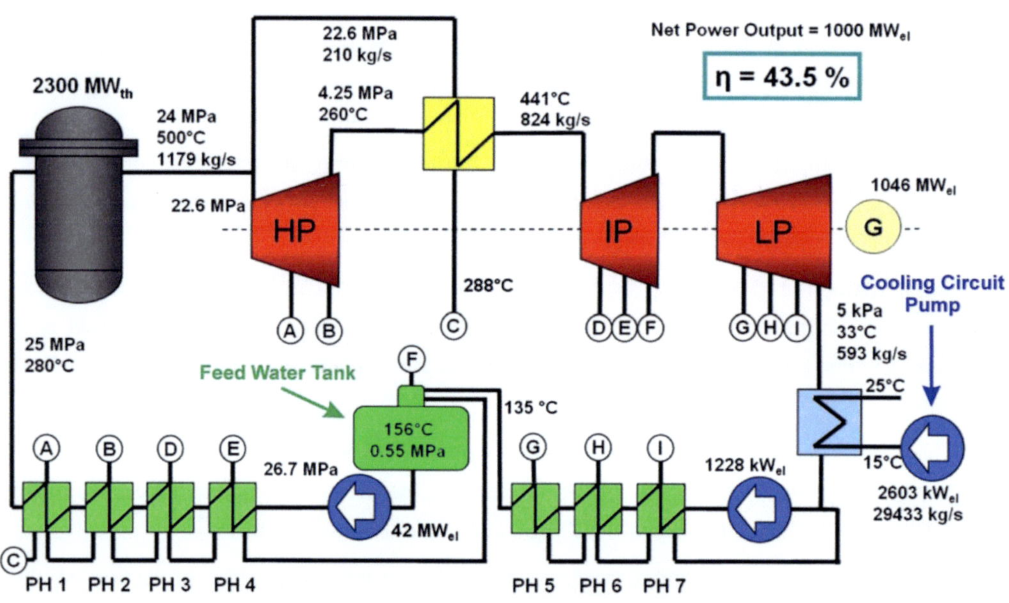

Fig. 5.1: Schematic diagram of the HPLWR steam cycle [6]

Like in Niederaussem, a high pressure (HP), medium pressure (IP) and low pressure (LP) turbine train is foreseen with a reheater between the HP and the IP turbine. Different from coal fired power plants, a steam fraction of 210 kg/s is extracted upstream the HP turbine to reheat the steam after the HP turbine. This extracted steam is cooled down to 288°C and thus changes its properties from steam like to liquid like conditions. Due to the non-linear cool down curve on the hot side, the IP steam on the secondary side can only reach a temperature of 441°C at maximum [6]. The reheat pressure of 4.25 MPa has been optimized such that condensation in the HP turbine will be avoided with some margin. Thus, a moisture separator as needed for the saturated steam cycle of boiling water reactors will not be required. The thermodynamic optimum of the cycle efficiency was at a slightly lower reheat pressure of 3.9 MPa, as discussed by Herbell et al. [7].

The process is shown in the temperature-entropy-diagram in Fig. 5.2 [6]. The HP turbine is expanding the steam such that the saturation line is almost reached. The steam quality at the outlet of the LP turbines of 85% will require moisture separation inside the LP turbines to avoid droplet erosion and to increase LP turbine efficiency, as will be discussed in Chapter 5.3.1.

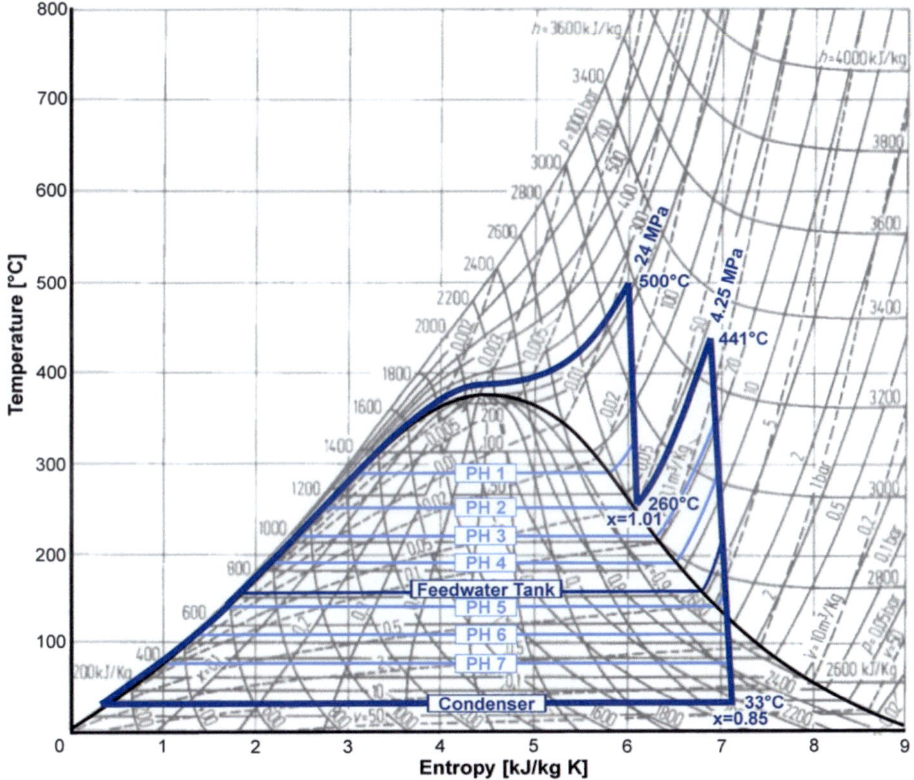

Fig. 5.2: The HPLWR steam cycle in the temperature-entropy-diagramm [6]

Oka et al. [1] as well as Bogolovskaya and Abdulkadyrov [8] discuss the alternative option to place the reheater between the IP and the LP turbines at 0.28 MPa. As the IP turbine will expand into the two-phase region in this case, they propose to use a moisture separator and two reheaters, one of which will be heated by steam extracted inside the HP turbine and a second one by life steam extracted upstream the HP turbine.

An interesting alternative has been proposed by Yamada et al. [9]. If sufficient moisture can be extracted inside the LP turbines, the reheater can completely be omitted and simply a moisture separator at 1.1 MPa, between the IP and the LP turbines, is considered to be sufficient. Such a solution, however, is strongly dependent on the IP and LP turbine design which has to stand the increased moisture level. Ideas to reheat the IP steam inside the reactor, as proposed for pressure tube reactors by Duffey et al. [10], are not considered to be applicable for pressure vessel type reactors as different pressure levels would require a significant mass of steel structures inside the reactor core.

From the choice of different options for reheating we can conclude that the advantages of a reheat process are not as obvious as with coal fired power plants, where the additional heat for the reheater is taken from the exhaust gas in the boiler and power as well as thermal efficiency increase significantly. With SCWR, instead, the heat for the reheater is taken from the main steam line which implies a loss of exergy of the live steam: if the extracted steam would pass all three turbines, it would produce more power than reheating the steam which is passing only the IP and LP turbines. Herbell et al. [[7], [11]] show, however, that the optimum reheat pressure can only be determined properly if also the flow losses by condensation inside the turbines are taken into account. For illustration, we compare in Fig. 5.3 the gross power of the turbine train as a function of reheat pressure and temperature after reheat without flow losses by condensation (left) with the gross turbine power including the flow losses by condensation inside the turbines (right). Without flow losses, the turbine power and thus the thermal efficiency would increase monotonically with the reheat pressure due to the decrease of exergy losses. Including these flow losses, the gross turbine power has a maximum when the steam at the outlet of the HP turbine reaches the saturation line. The chosen reheat pressure of 4.25 MPa [6] adds some margin from the saturation line.

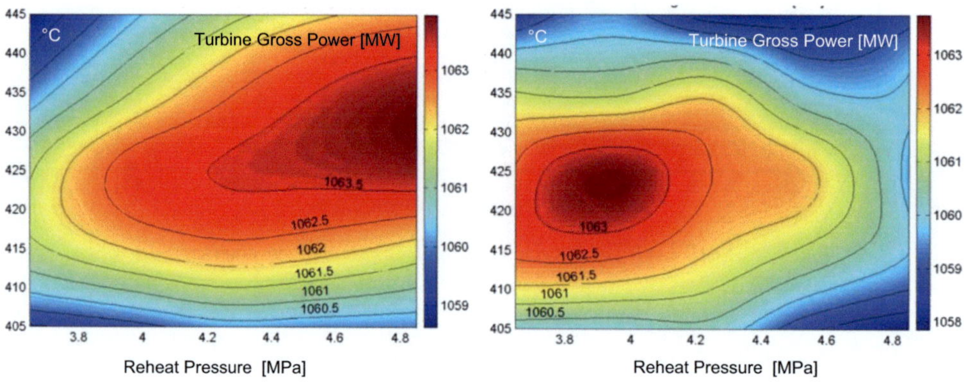

Fig. 5.3: Gross power of the turbine train as a function of reheat pressure and temperature, without consideration of flow losses by condensation inside turbines (left) and with flow losses (right) [11]

The feedwater line, shown in Fig. 5.1, includes 8 preheater stages of which the feedwater tank at 0.55 MPa serves as one of the preheaters. The high pressure condensate left over from the reheater is added to the final, high pressure preheaters PH1. Besides its function as a water reservoir and expansion vessel, the feedwater tank serves also as a deaerator. As an interesting alternative, Bogolovskaya and Abdulkadyrov [8] propose to design the feedwater line without any feedwater tank and to use an enlarged hotwell of the turbine condensers as feedwater reservoir and deaerator instead.

5.3 Major components of the steam cycle

A number of detailed design studies have been performed for selected components of the reference cycle described in Fig. 5.1, as will be discussed next.

5.3.1 Steam turbines

A suitable turbine design concept, shown in Fig. 5.4, has been discussed by Herbell et al. [12] based on state of the art steam turbines of fossil fired power plants. Like there, the turbine train may run at full speed of 50 Hz. The characteristic parameters of the HPLWR steam turbines are listed in Tab. 5.1. At the given condenser pressure of 5 kPa, the last stage of the LP turbine will need an axial exhaust area of 12.5 m^2. The three double flow LP steam turbines, with a bearing distance of 9.2 m and an estimated weight of 350 t each, are designed with 7 stages per flow. Also for the IP turbine, a comparable unit to those of supercritical fossil-fuelled power plants can be implemented. The proposed turbine is a double flow unit with asymmetric extractions at different pressures, including blade heights from 137 mm to 287 mm and a rotor with 1200 mm diameter at the last stages. Ten stages are needed for generator side (GS) flow, and nine stages are housed on the turbine side (TS).

Fig. 5.4: Turbine design concept with double flow HP, IP and 3 LP turbines [12]

It is only the HP turbine that exceeds the dimensions of current HP turbines with its mass flow of 970 kg/s and its power of 320 MW. Herbell et al. [12] propose to use a double flow

HP turbine having the advantage of axial thrust compensation. It would have to be designed particularly for this purpose, which is not expected to be a major challenge though. Steam inlet temperatures will even be around 100°C lower than those for supercritical fossil fired power plants. 17 turbine stages with a blade height of 48 mm at the inlet, increasing to 139 mm at the outlet could yield a rotor with 840 mm maximum diameter and a bearing distance of 6.5m.

The enthalpy and pressure characteristic of the HP turbine is shown in Fig. 5.5. The highest pressure and enthalpy drop occurs at the diagonal inlet stage. Such a diagonal stage leads the radial steam flow from the inlet most efficiently to the axial blading. The inlet size section is minimized and the risk of flow separation at the inner casing is reduced. The steam extraction is placed after the 12th stage. The HP turbine features a drum-type shrouded blade design concept. The blade and vane of the diagonal inlet stage as well as the blades and guide vanes of the first and second blading drum are made of X20Cr13, a ferritic-martensitic steel.

	HP	IP (GS/TS)	LP
Number of flows	2	2	6
Number of stages	17	10/9	7
Min. blading height	48 mm	137/137 mm	58 mm
Max. blading height	139 mm	287/282 mm	1145 mm
Max. rotor diameter	840 mm	1200 mm	1900 mm
Bearing distance (per unit)	6500 mm	7000 mm	9200 mm
Approx. weight (per unit)	180 t	200 t	350 t

Tab. 5.1: Characteristic parameters of the HPLWR steam turbines [12]

The low steam quality of steam of 85% at the outlet of the LP turbines requires a droplet separation inside the LP turbines to avoid droplet erosion. A higher exit steam quality of 87% can be achieved by separating droplets through the steam extractions as shown in the enthalpy-entropy-diagram, Fig. 5.6, at steam temperatures of 105°C and 64°C [7].

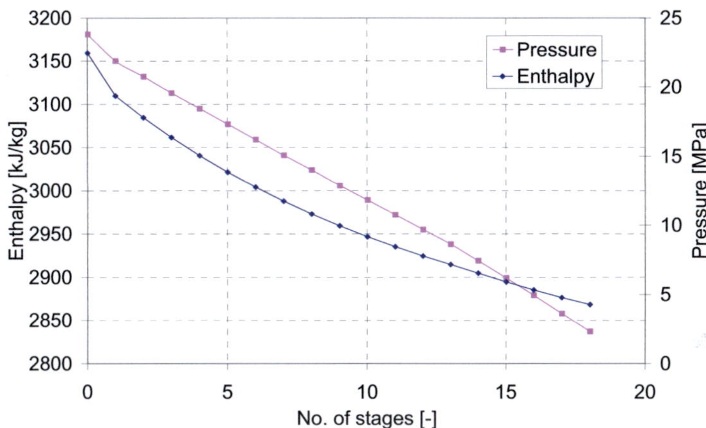

Fig. 5.5: Expansion line of steam pressure and enthalpy in the HP turbine [12]

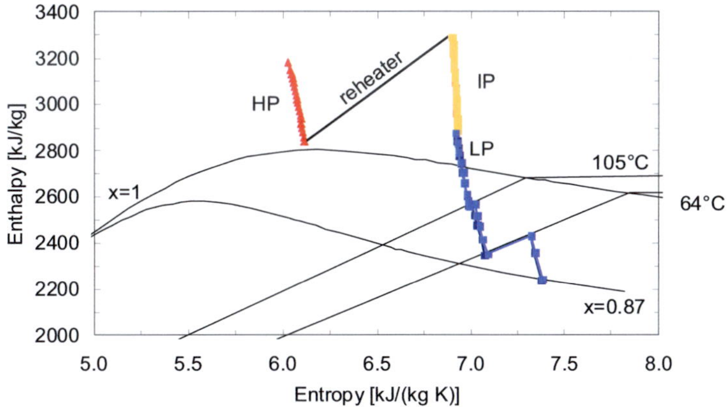

Fig. 5.6: Steam expansion in the HP, IP and LP turbines with moisture separation in the LP turbine [7]

5.3.2 Reheater

A vertical reheater design concept with straight tubes has been proposed for the HPLWR by Herbell et al. [12]. Two parallel reheaters with 20 m total axial length, 15 m tube length and an outer diameter of 3 m provide a heat transfer surface area of 6868 m^2 each. Due to the supercritical pressure on the hot side, the thickness of the tube sheet becomes the limiting parameter which should not exceed 800 mm because of manufacturing constraints and to limit transient thermal stresses. If arranged horizontally, this reheater should slightly decline to avoid parallel channel instabilities.

As indicated in Fig. 5.7, the axial temperature profile on the hot side is highly nonlinear, which reflects the transition from steam like to liquid like conditions at supercritical pressure, which could be described as pseudo-condensation. On one hand, this effect limits the achievable outlet temperature of the IP steam to 441°C at a given reactor outlet temperature of 500°C. On the other hand, it includes heat transfer effects on the high pressure side with highly non-linear fluid properties. Herbell et al. [13] studied such heat transfer in more detail using test data obtained with supercritical CO_2 as a reference. They show that a deterioration of heat transfer may occur also under cooling conditions due to an obvious laminar flow even at Reynolds numbers up to ~20,000. Even though such Reynolds numbers are outside the range of the proposed reheater concept, further experimental studies with supercritical water under cooling conditions are recommended to confirm this unexpected flow effect.

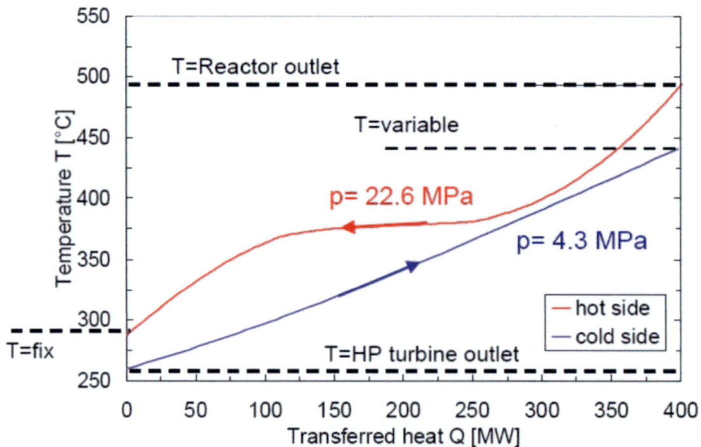

Fig. 5.7: Temperature vs. transferred heat of the reheater concept [7]

Dimensions of the reheater proposed by Herbell et al. [7] are listed in Tab. 5.2. The tube length of 15 m is in the range of commercially available heat exchanger tubes. To avoid thick

194

walled tubes due to the high pressure on the primary side, a small tube diameter of 16 mm has been preferred. The tubes shall be expanded and welded afterwards in the tube sheet. As the thickness of this support plate is limited to 800 mm, as described above, an arrangement with two parallel heat exchangers became necessary, reducing the tube sheet thickness to 500 mm.

Number of reheaters	2
Total heat transfer area (one unit)	6868 m²
Shell side	
Design pressure	5 MPa
Design temperature	450°C
Outer diameter	3.0 m
Wall thickness	0.08 m
Material	P 91
Tube side	
Design pressure	26 MPa
Design temperature	500°C
Number of tubes (One Unit)	9450
Tube length	15 m
Outer diameter/pitch	16/21 mm
Wall thickness	2 mm
Material	P 91

Tab. 5.2: Dimensions of the reheater [11]

The conceptual design of such a reheater is shown in Fig. 5.8. The lower plenum, collecting the high pressure liquid after pseudo-condensation inside the tubes, is mechanically separated from the shell, enabling free axial expansion of the tube bundle. Horizontal displacements of the lower plenum are avoided by guide rails of the shell, and piston rings at the sliding outlet nozzle of the lower plenum minimize leakage from the HP side to the IP side. The IP steam is running upwards with a counter current flow on the shell side as indicated in Fig. 5.7. A man hole to the upper and lower plenum each allows inspection of the tube sheet and its weldings. The overall length of 20 m of the proposed heat exchanger is comparable to reheaters operated in present pressurized water reactors, but the outer diameter of about 3 m is significantly smaller. Due to the higher pressure, however, the total weight of 325 t per heat exchanger is in a similar range. The reheater arrangement can be horizontal or vertical, depending on the chosen layout of the turbine building.

The high pressure side will experience a significant increase of coolant density as shown in Fig. 5.9, which might cause flow instabilities. Herbell et al. [14] analyzed static and dynamic flow instabilities like Ledinegg instabilities, parallel channel instabilities and dynamic instabilities for different orientations of the reheater, applying verified two-phase flow methods to the supercritical water.

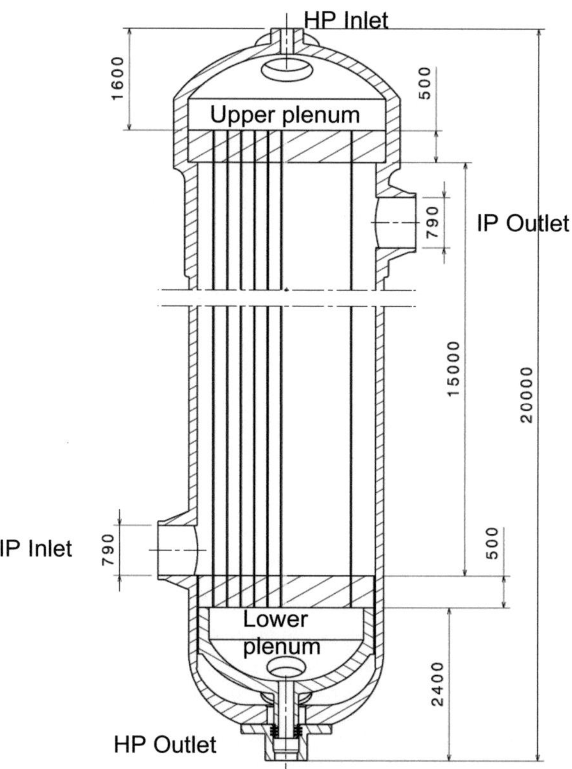

Fig. 5.8: Conceptual design of the reheater [11]

Parallel channel instabilities can occur if the reheater is inclined and if the HP mass flow is small, such that steam rises in some tubes and liquid flows back in other ones. This instability can be avoided if the HP outlet is lower than the inlet. Density wave oscillations cannot occur inside the tubes since the coolant density is smallest and the coolant pressure drop is largest at the inlets of the tubes, which stabilizes such instabilities. A choked flow instability as described by Herbell et al. [14] cannot occur either as the velocity of sound is much higher than the inlet steam velocity in this case.

196

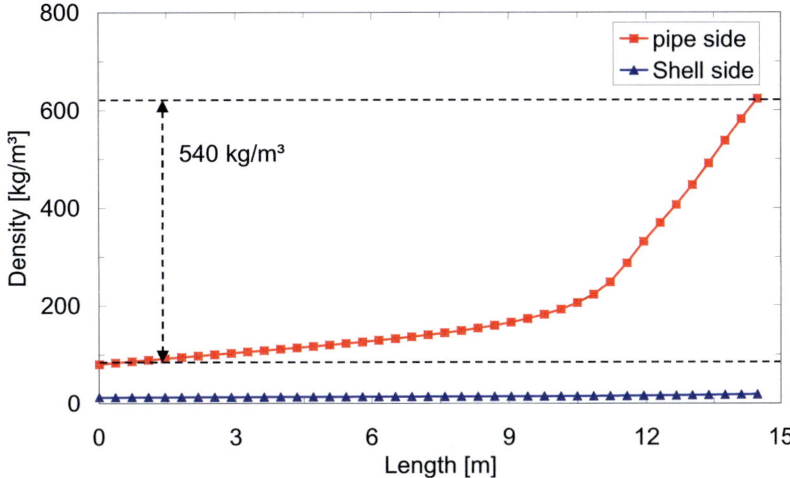

Fig. 5.9: Density change of HP steam on the tube side and IP steam on the shell side of reheater tubes [14]

5.3.3 Feedwater pumps

A set of 4 parallel feedwater pumps are recommended for the HPLWR concept, of which 3 are providing the full load mass flow rate of 1179 kg/s and a number 4 unit is kept on hot stand-by to continue operation in case of a pump trip. They shall provide a feedwater pressure of 26.7 MPa under nominal conditions. High pressure, multistage centrifugal pumps designed for supercritical fossil fired power plants can be used for this purpose, in principle, except that this nuclear application will require a leak tight enclosure. Type CHTD high pressure barrel casing pumps of KSB [15] are typical examples which could be qualified for the HPLWR.

Note that the feedwater pumps are the only coolant supply for the core. There is no primary or recirculation pump foreseen in the HPLWR and a simultaneous trip of all coolant pumps would cause a loss of coolant in the core. In this case, the reactor needs to be depressurized to maintain the flow rate, either by blow down of steam to the condensers or, under accidental conditions, to the depressurization chamber of the containment, as discussed in Chapter 4. An extended coast down time of the pumps, e.g. provided by flywheels, does not help in these cases. The parallel feedwater pumps need to be equipped with check valves to avoid backflow in case of a trip of a single pump, and the feedwater penetration through the containment is closing passively by check valves as well. These check valves will make the pumps ineffective during coast down.

5.3.4 Feedwater tank

The feedwater tank for the HPLWR steam cycle has been designed by Lemasson [16] for a maximum water volume of 350 m^3 at a nominal pressure of 0.55 MPa. Steam which has been extracted near the outlet of the IP turbine (with 0.75 MPa, 233°C, 19.7 kg/s) is released here through spargers in a horizontal tube inside this tank. The feedwater line (with 1.25 MPa, 165°C. 438 kg/s) as well as the condensate lines from high pressure preheaters (with 0.75 MPa, 135°C, 721 kg/s) are supplying water to the tank under all load conditions (data at full load). During the start-up phase, the start-up system described in chapter 5.4 and condensate bypass lines from both preheaters PH1 are supplying additional water and are thus heating up the water reservoir in the tank. A vent pipe connected with the tank serves as a deaerator to remove gas arising from the steam extractions. Finally, the feedwater tank controls the inlet pressure head of the feedwater pumps to avoid a booster pump. Fig. 5.10 shows an example how this tank could be designed.

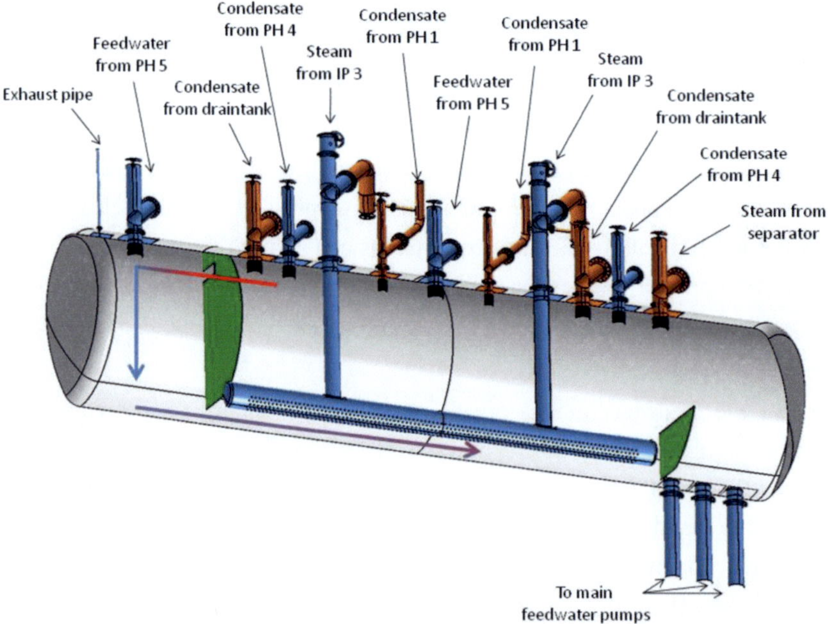

Fig. 5.10: Conceptual design of the feedwater tank with up to 350 m^3 of feedwater at 0.55 MPa [16]

The tank is divided into 3 compartments which are connected by penetration holes. This minimizes droplet entrainment from the spargers into the vent line or bubble entrainment into the supply lines of the feedwater pumps but still provides a cross flow, and thus shall provide sufficient coolant mixing to keep the entire water volume at saturation temperature. A co-axial feedwater line, coming from the upstream preheater PH5 is cooling the exhaust pipe to condense steam extracted with the non-condensable gas.

5.3.5 Feedwater preheaters

Two lines of preheaters, with 7 U-tube heat exchangers each, have been proposed for the HPLWR plant concept [17]. The high pressure preheaters, PH1 to PH4 in Fig. 5.1, have an outer diameter of 2 m to 2.5 m and a length of up to 12.3 m, of which PH1 is the largest component. Fig. 5.11 shows a cut away view of PH1. Steam which has been extracted from the steam turbines is entering the tube bundle from the top. As this steam is still superheated, is has to be cooled down first to the saturation temperature in the upper quarter of the preheater. A steam jet impinging on the tube bundle should be avoided because of the risk of corrosion-erosion. Condensate from upstream preheaters or from the reheater is entering from the right and evaporating again at the lower pressure. The steam is condensing on the tube outer surfaces and is finally sub-cooled in a compartment in the lower half of the preheater. Like with the reheater, the limiting component of this heat exchanger is the tube sheet which should not exceed 800 mm thickness. The tube weldings can be inspected from the inlet and outlet plenums through a man-hole. These components are quite similar to those in fossil fired power plants so that major challenges are not expected.

Fig. 5.12 shows a comparison of the outer dimensions of a high pressure and a low pressure preheater. The preheaters have been designed and dimensioned by Brandauer [18]. He assumed 2 lines of 7 preheaters each, of which 4 high pressure preheaters (PH1 to PH4) are placed downstream the feedwater pumps and 3 low pressure preheaters (PH 5 to PH 7) are downstream the condensate pumps. PH7 may also be plugged into the condensers to minimize the low pressure steam extraction lines. A high pressure preheater, shown in Figure 5.12 (top) has an outer diameter of 2.5 m with a total length of 12.3 m. A low pressure preheater, shown exemplarily in Figure 5.12 (bottom), has an outer diameter of 1.7 m with a total length of 11.9 m. More characteristic data can be found in Tab. 5.3, taken from Brandauer [18]. With a total weight of 1088 t of both feedwater lines, the preheaters contribute certainly a significant part to the total costs of the steam cycle. Alternative solutions with a smaller number of preheaters, with a compromise in thermal efficiency, could be considered in case of low fuel costs.

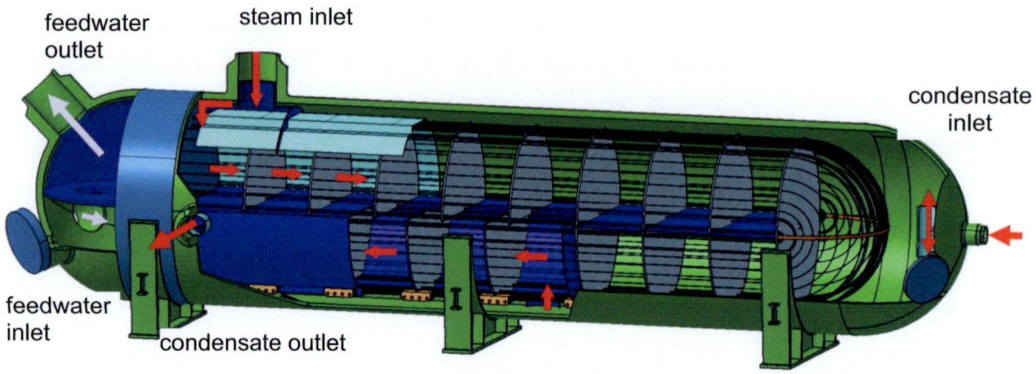

Fig. 5.11: Design concept of the high pressure preheater PH1 [17]

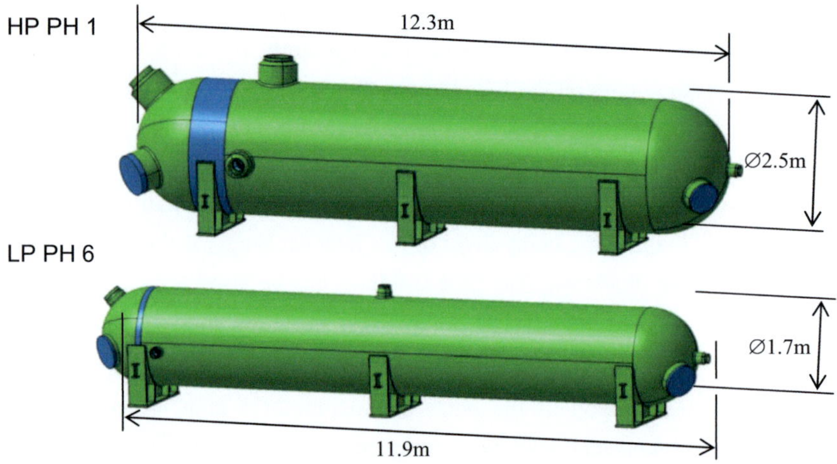

Fig. 5.12: Size comparison of a high pressure (top) and a low pressure (bottom) preheater [17]

Deformations and stresses of the thick-walled tube sheet of preheater PH1 have been analyzed by Schlageter [19] with the finite element code ANSYS under steady state conditions. The tube sheet is mechanically loaded with the feedwater pressure of 26 MPa on the colder side, maximum temperature 280°C, vs. 6.82 MPa and 320°C on the steam side. Two times 3600 tube penetrations with a diameter of 16 mm and a pitch of 21 mm in a square arrangement are weakening the plate with its 2.42 m diameter and 684 mm thickness. Local stress and deformation analyses of single penetrations have been performed to homogenize the weakened part of the plate with orthotropic elements of lower stiffness. The modeled geometry is shown in Fig. 5.13 with the perforated part shown in green. The spherical shell (brown), the separation plate (gray) and the cylindrical shell (blue) are stiffening the plate, causing peak stresses on the other hand which need to be minimized by large radii.

Preheater	PH1	PH2	PH3	PH4	PH5	PH6	PH7
Steam extraction no.	A	B	D	E	G	H	I
Steam pressure [MPa]	6.82	3.95	2.26	1.31	0.36	0.13	0.04
Steam temperature [°C]	320	256	365	299	159	108	76
Steam mass flow [kg/s]	44.8	27.8	20.9	20.4	21.4	19.6	23.1
Outer diameter [m]	2.488	1.954	2.238	2.272	1.730	1.663	1.934
Total length [m]	12.3	12.1	12.4	12.3	12.0	11.9	12.15
Number of U-tubes	3600	2116	3025	3136	1600	1444	2116
Thickness tube sheet [mm]	684	547	639	652	88	93	119
Total weight [t]	133	73	94	98	45	43	58

Tab. 5.3: Characteristic data of the HPLWR preheaters [18]

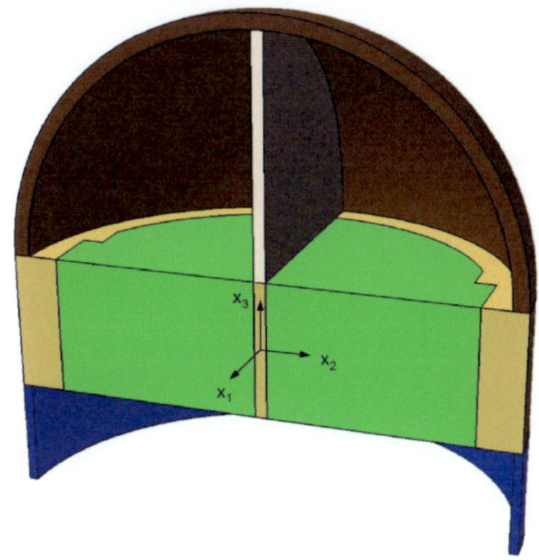

Fig. 5.13: Finite element analysis of the tube sheet of the HP preheater PH1 [19]

Fig. 5.14, left, shows exemplarily the global stress distribution in lateral direction in the homogenized part of the plate, with a maximum stress of 54 MPa. The stress concentration factor of the penetrations in this case is up to 5.1, shown in Fig. 5.14, right, resulting in a concentrated stress of up to 275 MPa which must be assessed against low cycle fatigue.

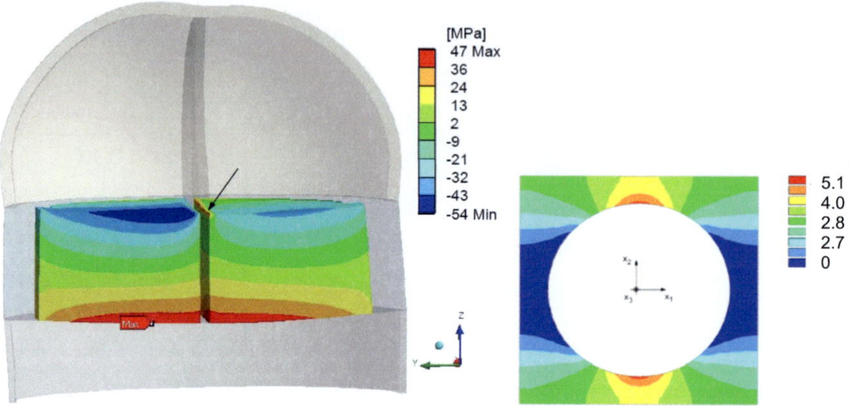

Fig. 5.14: Deformations (80x magnified) and lateral stresses (x-direction) of the homogenized, perforated part of the tube sheet (left) and stress concentration factor of a single tube penetration under lateral load (right) [19]

Concentrated stresses in the transition radius from the central separation plate next to a tube penetration, indicated with an arrow in Fig. 5.14, are shown exemplarily in Fig. 5.15. The peak stresses are reaching up to 369 MPa there, which is already close to the fatigue limit of the chosen material 20MnMoNi5-5.

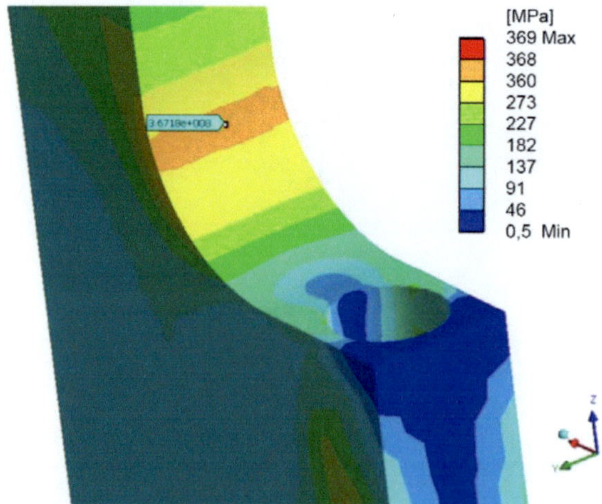

Fig. 5.15: Peak stresses in the transition radius from the central separation plate to a tube penetration [19]

5.4 Control of the steam cycle

For discussion of the control of the steam cycle, we need to differ between the load range, when the turbine generator is connected with the grid, and a start-up range, when the system is pressurized, the thermal power of the reactor core is ramped up to a minimum power, and the components of the steam cycle are warmed up.

5.4.1 Control in the load range

Following a proposal of Schlagenhaufer et al. [20], the HPLWR should have load following capabilities at least between 50% and 100% of the nominal thermal power of the core. In this range, they propose to keep a constant, supercritical feedwater pressure of 25 MPa at the reactor inlet by control of the turbine valve. The steam temperature should be kept at the nominal temperature of 500°C at the reactor outlet to gain high part load efficiency, either by controlling the core power by the control rods and the steam temperature by the speed of the feedwater pump, as suggested by Schlagenhaufer et al. [20], or vice versa by controlling the core power by the feedwater pumps and the steam temperature by the control rods of the core. The latter option has been preferred by Oka et al. [1].

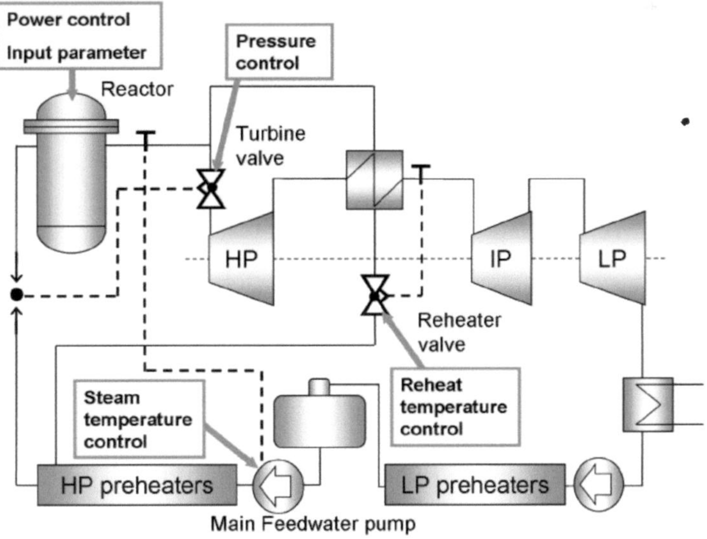

Fig. 5.16: Control loops for operation in the load range [20]

The reheat steam temperature at the inlet of the IP turbine can be kept at 441°C by a valve in the high pressure line downstream of the reheater, controlling the mass flow rate being extracted from the life steam. Fig. 5.16 shows these control loops in a simplified scheme.

The control system has been modelled by Schlagenhaufer et al. [20] with the system code APROS [21] to check its performance. Reactivity control has not yet been included in these analyses. Several additional control loops had to be added like liquid level control of the condensers by the condenser pump, condensate level control in the preheaters by their drain valves, and outlet temperature control of the cooling loop by the cooling pumps. No control has been foreseen, however, for the feedwater temperature at the reactor inlet and for the reheat pressure. Deloading the reactor from 100% to 50% of its maximum power in a step function, as indicated in Fig. 5.17, we can demonstrate the system response in Figs. 5.18 and 5.19.

Fig. 5.18 (left) shows that the pressure at reactor inlet is kept constant by the turbine valve, while the outlet pressure is increasing with decreasing load because of the smaller pressure drop at lower coolant mass flow rate. A small and stable pressure fluctuation at the reactor inlet is due to the compressibility of the steam. The reactor outlet temperature is kept constant by control of the feedwater pump, as shown in Fig. 5.18 (right), which yields a high net efficiency of more than 41% even at 50% load. The reactor inlet temperature is decreasing, however, from 280°C to 245°C when deloading to 50%. The reason is a decreasing pressure of the steam extractions, Fig. 5.19 (left) and thus a decreasing temperature of the extractions, in particular of the high pressure extractions A and B, Fig. 5.19 (right), according to the elliptic law of steam turbines. The effect of this feedwater temperature swing on reactivity should be checked carefully in future studies.

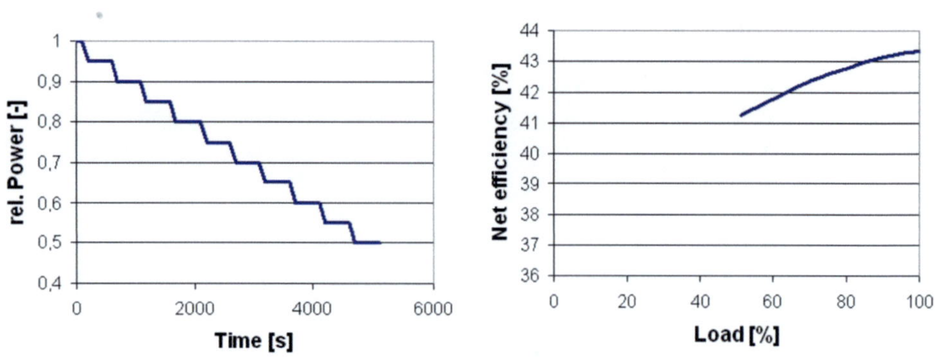

Fig. 5.17: Stepwise power reduction (left) and corresponding net efficiency in the load range [20]

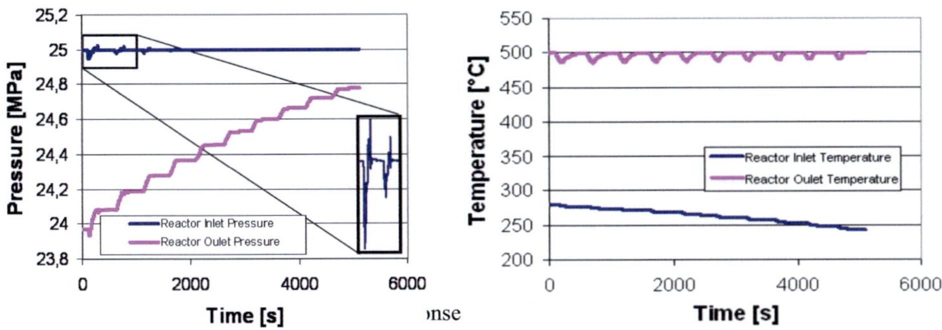

Fig. 5.18: Pressure and temperature response at reactor inlet and outlet [20]

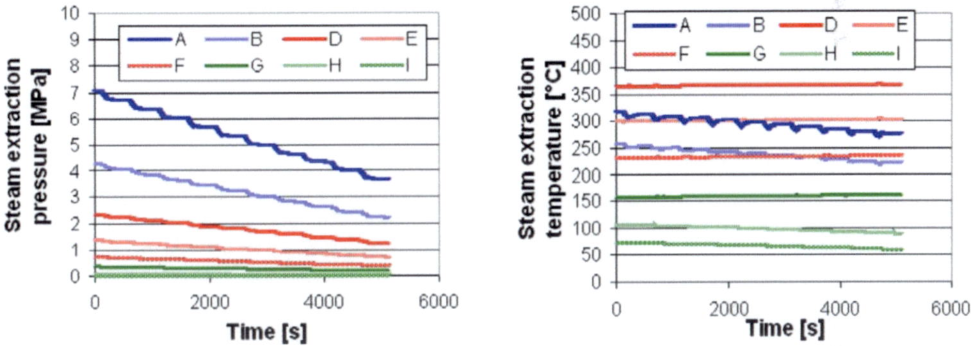

Fig. 5.19: Pressure and temperature response of steam extractions [20]; symbols A to I refer to steam extractions as defined in Fig. 5.1

5.4.2 Start-up and shut-down procedure

At 50% load, the feedwater mass flow rate has been reduced to around 50% of its nominal value and the reactor pressure drop to around 25% only. Further reduction of the mass flow rate would risk that the moderator flow through some of the water boxes, Fig. 2.1, would start to reverse. Moreover, the downward flow through the first superheater requires a certain minimum mass flow rate to avoid local flow reversal. Therefore, the feedwater mass flow rate is kept constant at less than 50% load and the core outlet temperature decreases with decreasing load. The turbines must be disconnected in this load range to avoid droplet erosion.

On the other hand, trying to avoid two-phase flow for any thermal power of the reactor, except for residual heat removal, Schlagenhaufer et al. [22] propose to pressurize the reactor to 25 MPa already under cold conditions. Once the reactor became critical, it produces at first only hot water (liquid) which cannot be supplied to the turbine. Therefore, water or steam from the reactor is considered to by-pass the turbines for any thermal power below 50% of the nominal power. A combined shut-down and start-up system, sketched in Fig. 5.20, has been proposed which allows to preheat the feedwater system and the turbines during start-up and to cool down all thick walled components slowly during shut-down. The system consists of a high pressure control valve which takes over the task of pressure control from the turbine valve, followed by a battery of separators, followed by a drain tank to provide hot steam and liquid independently to preheat the feedwater tank, the preheaters and the turbines. Excess steam, in particular during the shut-down phase, is supplied to the condensers. The system is controlled by

- Loop α which controls the pressure control valve such that the reactor inlet pressure is kept at 25 MPa.

- Loop β which controls the feedwater pump such that the feedwater mass flow rate is kept at 50% of the full load value

- Loops γ and δ which control the temperature of the feedwater tank either by extracting steam from the separator if the feedwater is too cold or by supplying liquid from the drain tank to the condenser if the feedwater is too warm.

- Loop ε which controls the feedwater temperature at reactor inlet to 241°C using steam from the separators.

- Loop ζ which controls the liquid level in the drain tank.

The system was tested by Schlagenhaufer [22] with a numerical APROS simulation and showed to work successfully after some optimization. The feedwater mass flow was kept constant up to 50% thermal power, so that the outlet enthalpy of the reactor was ramped up proportionally to the thermal power. At 50% power, a steam temperature of 500°C was reached and the system could be switched over to the turbines.

Velluet [23] and Schlagenhaufer [24] proposed to use a battery of 4x24 smaller cyclones of 410 mm diameter each to separate a two-phase flow of 600 kg/s at maximum with a minimum of steel mass of 27.6 t only. A battery of 24 cyclones is sketched in Fig. 5.21. The two-phase flow enters the cyclone tangentially, causing a swirl which separates liquid at the lower outlet from steam at the upper outlet. The drain tank simply collects the liquid phase from the separators, shown in Fig. 5.21 (right).

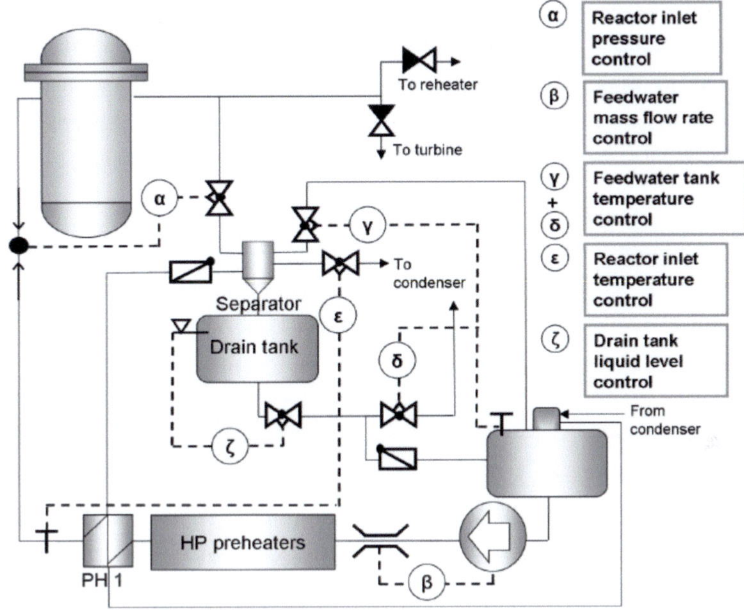

Fig. 5.20: Control of the combined start-up and shut down system [22]

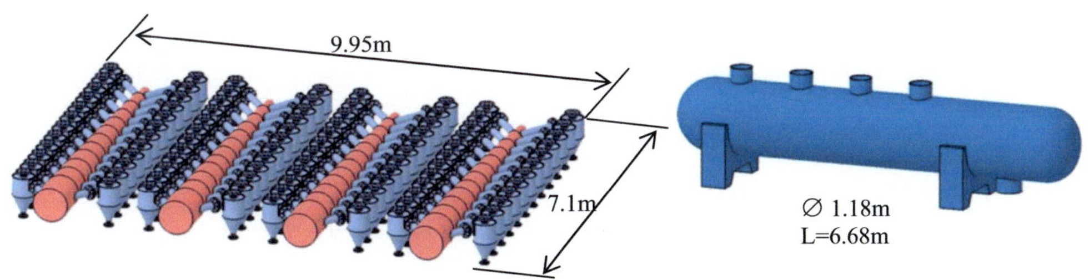

Fig. 5.21: Cyclone battery separating the two-phase flow in the combined start-up and shut down system (left) and drain tank (right) [17]

To simulate the shut-down operation below 50% load, Schlagenhaufer [24] assumed in his APROS analyses that the thermal power of the reactor core follows a predefined operation curve, which is sketched in Fig. 5.22. The control rods were inserted into the core in 3 phases with 1000 sec each, starting with superheater 2 and followed by superheater 1, before the evaporator control rods are finally inserted. The APROS simulation changed to decay heat in each heating zone once 6% thermal power had been reached. The reason for the stepwise shut-down of the reactor core is that the transition through the pseudo-critical temperature should stay in the evaporator. Having liquid and steam like conditions in superheater 1 could cause flow stability problems with downward flow of fluid with higher density and upward flow of fluid with lower density. A combined neutronic and thermal-hydraulic core analysis will be necessary, however, to confirm if such shut-down procedure can be realized in the core.

As mentioned before, the temperatures of the reactor inlet and of the feedwater tank are kept constant during shut-down as long as possible to minimize thermal stresses of the reactor pressure vessel. Therefore, the temperatures of the feedwater tank and of the reactor inlet were taken as setpoint values for the controllers to maintain initial values. At 50% load, the temperatures were 135°C in the feedwater tank and 241°C at the reactor inlet. When decay heat was reached in the core, the system was tripped through stop valves and the steam cycle was depressurized through the turbine bypass valve. To start-up the steam cycle, the shut-down procedure is reversed. The mass flow rate is set to the 50% load value. Nuclear heating can be started immediately after the system is pressurized to 25 MPa. To allow a smooth pressurization of the system, the valve α may not be closed completely but should always be kept open, at least by 0.1%. This can be realized technically in two different ways, either by having a valve which cannot be closed completely or by a certain bypass of the valve with small resistor tubes, which always allow a certain minimum mass flow.

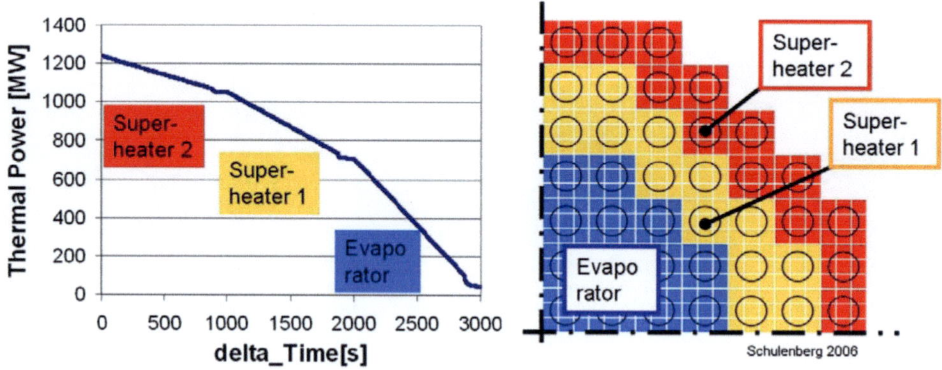

Fig. 5.22: Assumed shut down procedure of the core power [22]

The APROS simulation of a shut-down procedure is shown in Figs. 5.23 to 5.25. Starting from normal operation with 50% load, with the turbines still being connected, the shut-down process is initiated. Fig. 5.23 depicts the devolution of thermal power, of reactor inlet pressure and of the separator pressure. After 110 sec, the thermal power decreases according to the pre-described shut-down curve as shown in Fig. 5.22 and reaches the decay heat level in the total core after 3000 sec.

Due to the line switch from normal operation mode to the shut-down system, the reactor inlet pressure and the separator pressure show a pressure overshot of less than 1 MPa and 2.5 MPa respectively, until a constant value is reached, which stays constant until the whole system is depressurized through the turbine bypass after 3110 sec. Fig. 5.24 shows the history of temperatures at the reactor inlet, the reactor outlet, the separator, and the feedwater tank. The reactor outlet temperature is slightly raised by 25°C during the line switch after 10 sec, since the main steam temperature is not controlled any longer. After that, it decreases due to the decreasing thermal power and reaches 252°C before the system is depressurized through the turbine bypass. The separator temperature follows the reactor outlet temperature immediately and reaches the the reactor inlet temperature after 1250 sec. The reactor inlet temperature and feedwater tank temperature stay constant during the shut-down procedure, since these temperatures are controlled by the system. After 2900 sec, however, the reactor inlet temperature has to decrease because the control system cannot keep the temperature constant any longer due to the further decreasing reactor outlet temperature. After depressurization, all temperatures decrease because the whole steam cycle is flooded with cold condenser feedwater and no temperature peaks are observed in the core.

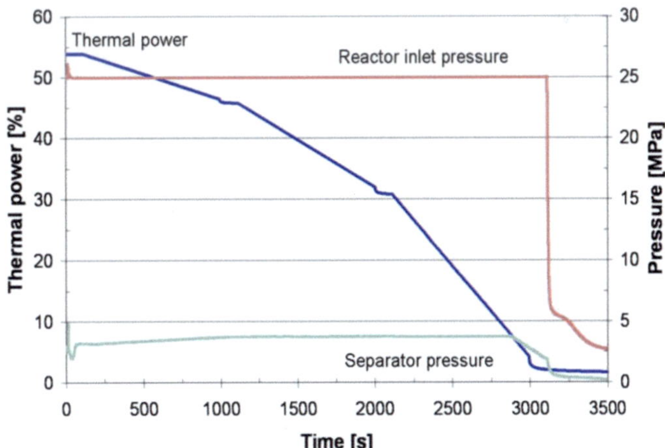

Fig. 5.23 Thermal power, reactor inlet pressure and separator pressure during the shut-down procedure [22]

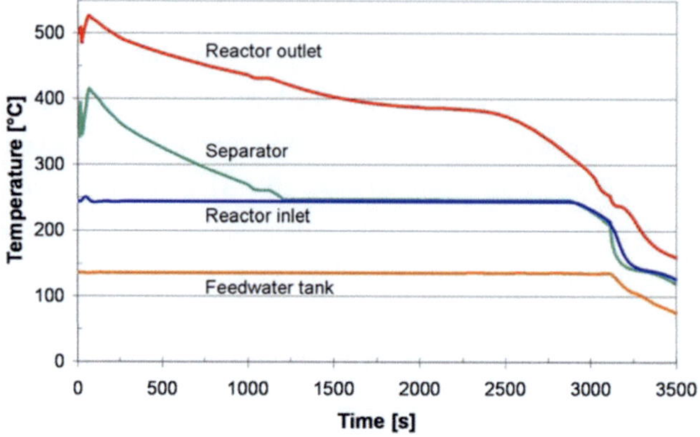

Fig. 5.24: Coolant temperature at reactor inlet, reactor outlet, in the feedwater tank and in the separator during the shut-down procedure [22]

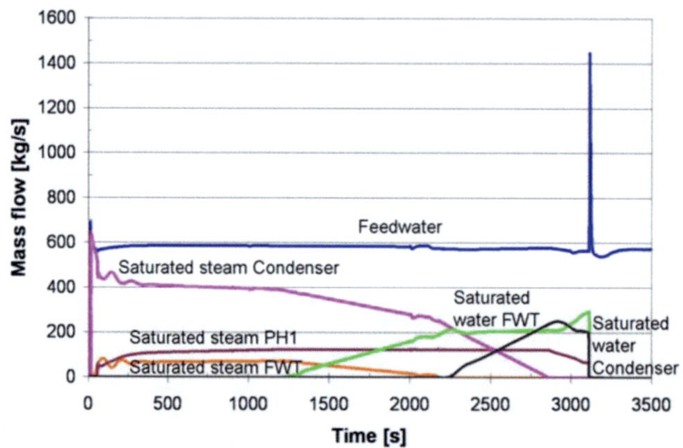

Fig. 5.25: History of mass flows of the shut-down system [22]

The history of different mass flow rates is depicted in Fig. 5.25. The turbine and the reheater are tripped and all steam is supplied to the separator, where it is distributed to PH1, the feedwater tank and the condenser, respectively. At first, as long as no water is separated, only steam is used to preheat the feedwater in PH1 and the feedwater tank, although ~ 75% of the steam is dumped into the condenser. Moreover, no steam is led to PH1 in the first 45 sec after activation of the shut-down operation, because the pressure of the PH1 is higher than the separator pressure during that time and the check valve is closed. Saturated water is produced after 1200 sec and led to the feedwater tank, since the fluid is throttled into the two-phase region, as the reactor outlet temperature drops below ~ 425°C, which is the intersection temperature of the 25 MPa - isobar and the condensing line - isenthalp. The mass flow rate of saturated steam led to the condenser and to the feedwater tank decreases during the shut-down procedure, whereas the mass flow rate of saturated steam led to PH1 stays rather constant. After 2200 sec, the saturated water mass flow reaches an adequate value to maintain the temperature of the feedwater tank and the control system closes the saturated steam line of the feedwater tank. Now, the entire saturated steam is only used to heat-up the feedwater flowing through PH1. Since the reactor outlet temperature decreases further, even less saturated steam is separated in the separator and no steam is dumped into the condenser anymore after 2800 sec. After 3110 sec, the system is tripped and the corresponding mass flow rates decrease to zero. A short feedwater mass flow peak during line switch can be observed when the turbine bypass valve to the condenser is opened.

The system is started again 390 sec after depressurization. The start-up signal initiates the control system. Fig. 5.26 shows the increase of the thermal power, of the reactor inlet pressure and of the separator pressure. Nuclear heating starts 100 sec after the system is pressurized to 25 MPa with the start-up system and follows the pre-defined curve as defined in Fig. 5.22. A continuous pressurization of the steam cycle is observed and the reactor inlet pressure remains constant during the start-up procedure. The separator pressure increases to ~ 4 MPa and stays quite constant.

After 9700 sec, the start-up system is closed and the steam is led to the turbine and the reheater instead, which causes almost no reactor inlet pressure change during the line switch, whereas the separator pressure decreases somewhat. The increase of temperatures at the reactor inlet, at the reactor outlet, in the separator, and in the feedwater tank during start-up is shown in Fig. 5.27. The reactor outlet temperature increases according to the thermal power characteristics. The separator temperature, the reactor inlet temperature and the feedwater tank temperature follows the reactor outlet temperature immediately. After 4200 sec, the reactor inlet temperature and the feedwater tank temperature reach the setpoint value of the control system (241°C and 135°C respectively) and stay constant during start-up operation afterwards. The separator temperature, on the other hand, increases further to a maximum value of 375°C. Small undershoots of reactor inlet and outlet temperatures are observed during the line switch after 9600 sec.

The history of different mass flow rates during start-up are depicted in Fig. 5.28. The entire saturated water and steam is led to the feedwater tank and to PH1, respectively, during the first 700 sec after start-up. This means that no steam or water is dumped to the condenser and therefore a short-cut through the feedwater tank is caused until the setpoint temperatures of the feedwater tank and reactor inlet temperatures are reached. Water and steam which are not needed for heat-up are dumped into the condenser after 4200 sec. After 4600 sec, all the saturated water is supplied again to the feedwater tank and additional saturated steam is used to hold the temperature of the feedwater tank constant, since less and less water is separated due to the increasing reactor outlet temperature. No water is separated at all after the reactor outlet temperature exceeded 425°C after 6450 sec, and only steam is used to heat-up the feedwater in the feedwater tank. The line switch to normal operation through reheater and turbines is foreseen after the thermal power reaches 50% load. A steady state operation of the steam cycle is reached after 9600 sec. An oscillation of turbine and reheater mass flow is observed for 100 sec with a frequency of 0.5Hz and amplitude of +/-50 kg/s after the line switch. The reason is that the reheater valve, which controls the reheat temperature, is situated upstream of the hot reheater side. Small changes of reheater valve position cause large mass flow changes to the reheater, since steam is flowing into the reheater and condenses there, which causes large density changes until a stable operation is reached.

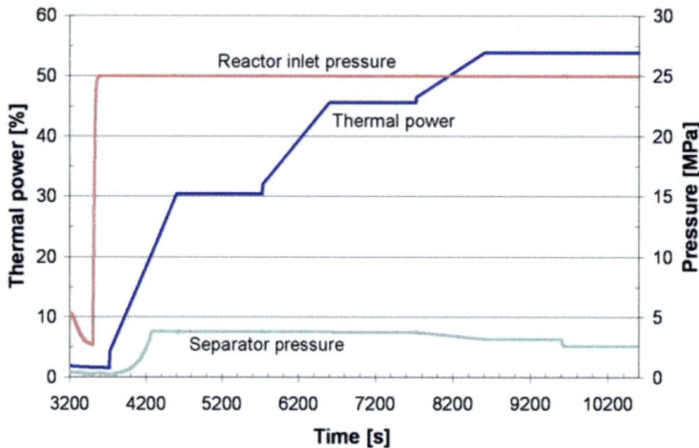

Fig. 5.26: Thermal power, reactor inlet pressure and separator pressure during the start-up procedure [22]

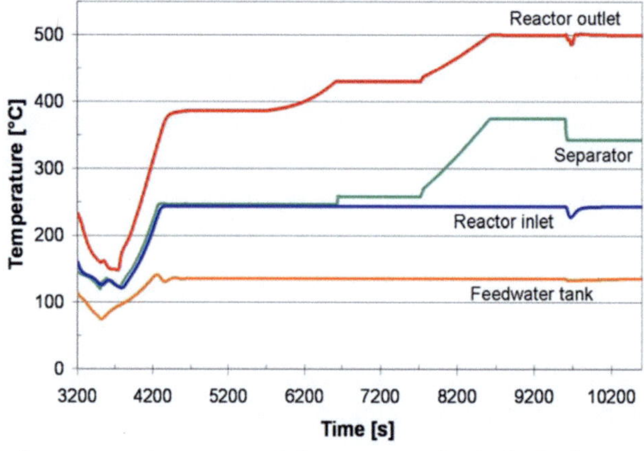

Fig. 5.27: Coolant temperature at reactor inlet, reactor outlet, in the feedwater tank and in the separator during the start-up procedure [22]

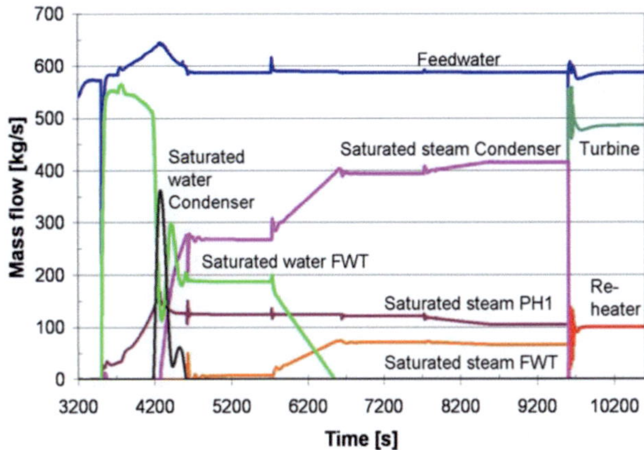

Fig. 5.28: History of mass flows through the start-up system [22]

Even though this procedure has still room for optimization, these examples demonstrate that reactor can indeed be operated at supercritical pressure of 25 MPa in the entire load range from cold water to full load conditions.

An alternative option to start-up and shut-down the reactor under full pressure has been proposed by Oka et al. [1]. They combined the tasks of the separator and of the drain tank in Fig. 5.20 to a single flash tank. It must be designed large enough to separate steam and liquid such that steam can be extracted from there without droplet entrainment to warm up the turbines. Oka et al. [1] predict a required flash tank height of 7.5 m and an inner diameter of 3.4m for a pressure of 6.9 MPa, resulting in a total weight of 52.3t for a plant size comparable to the HPLWR.

While a sliding pressure operation in the load range must be avoided to exclude dryout in the core, a sliding pressure start-up process with two-phase flow from the core outlet to the separator has been proposed alternatively by Oka et al. [1]. However, such system has not been considered as feasible for the HPLWR concept because of downward flow of coolant in the 1st superheater with the risk of flow reversal of steam.

A further optimization of the start-up system could be to combine the idea of the flash tank with the concept of the separator battery by including cyclones inside the flash tank, such that the tank size can be reduced at given maximum moisture content in the steam line to the turbines.

5.5　Layout of the power plant

The plant layout of the HPLWR presented here is based in principle on an Advanced Boiling Water Reactor like NPP Gundremmigen in Germany. A first layout of the HPLWR plant has been designed by Bittermann et al. [25] which has further been extended here to include details of the start-up system, the reheaters and the feedwater tank, as described above, as well as major piping. The isometric view, Figure 5.29, shows the HPLWR reactor building on the left and the turbine building on the right, connected with 4 feedwater lines (blue) and 4 steam lines (pink).

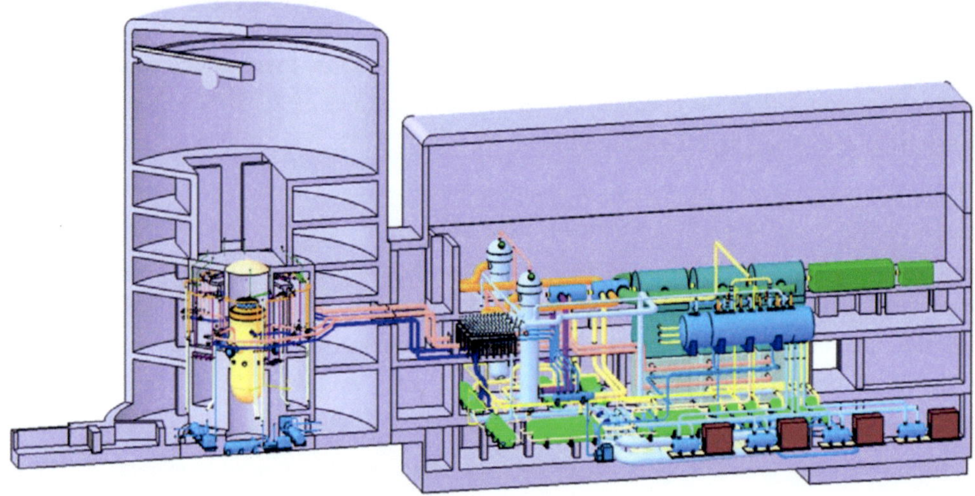

Fig. 5.29: Cut away view of the reactor building (left) and the turbine building (right) [17]

The reactor building of the HPLWR includes the containment with the reactor pressure vessel, the safety systems and the residual heat removal systems as described in Chapter 4.2. Additional compartments surrounding it are containing safety-related mechanical, electrical, instrumentation and control components and auxiliary systems, like the boron injection system, the reactor water and fuel pool cleanup system, the fuel pool cooling system and the radioactive liquid waste storage system. On top of the containment, the fuel handling area with the shielding and spent fuel storage pool can be found. The main function of the reactor building is to protect all safety-related equipment against the effects of natural and external man-made hazards. The reactor building as a secondary containment also guarantees confinement of radioactivity as the last barrier preventing the release of radioactive materials

214

to the outside atmosphere upon occurrence of a beyond-design event. Therefore, all components with a high radioactive inventory are placed within the reactor building. This includes also activated charcoal filters, the air recirculation system and the transport way for the spent fuel cask. Containment isolation valves prevent release of a significant radioactive inventory into the turbine building in case of a break. The main parameters of the reactor building are an outer diameter of 45m, an approximate height of 61m and an approximately volume of 95000m³. The total height is determined by the height of the containment, the height of the fuel pool on top, and the maximum required lift of the reactor building crane. Figure 5.30 shows a sectional view of the reactor building with containment, reactor pressure vessel and safety systems.

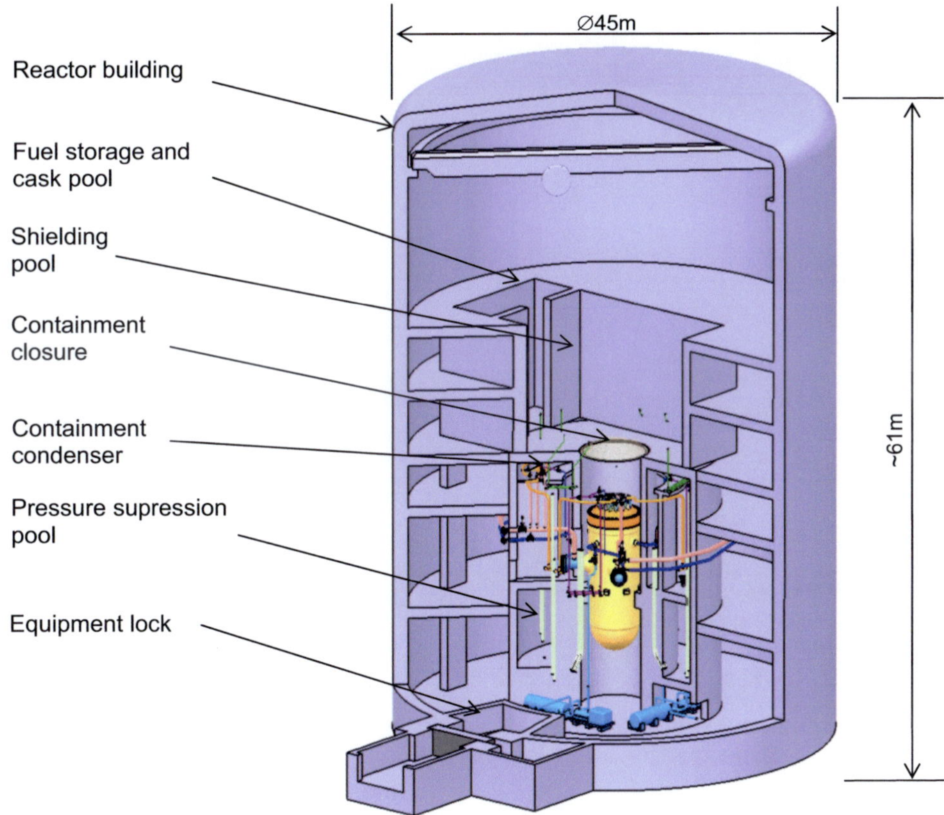

Fig. 5.30: Cut away view of the reactor building [17]

The turbine building is located next to the reactor building. Figure 5.31 shows a cut away view through the turbine building with steam lines and turbine bypass lines in pink, feedwater lines in blue, and steam extraction lines in yellow. The turbine building contains all components of the steam cycle like turbines, condensers, feedwater pumps, preheaters, reheaters and the start-up system. As the steam is activated, the turbine building is part of the controlled area of the plant. The main parameters of the reactor building are a length of app. 87 m, a width of app. 49 m, an app. height of 52 m, resulting in a volume of approximately 222000 m³. The length of the building is mainly determined by the turbine-generator set, while the width is determined by the workspace for the low pressure turbines including the condenser withdrawal length and by the preheater and pump arrangement.

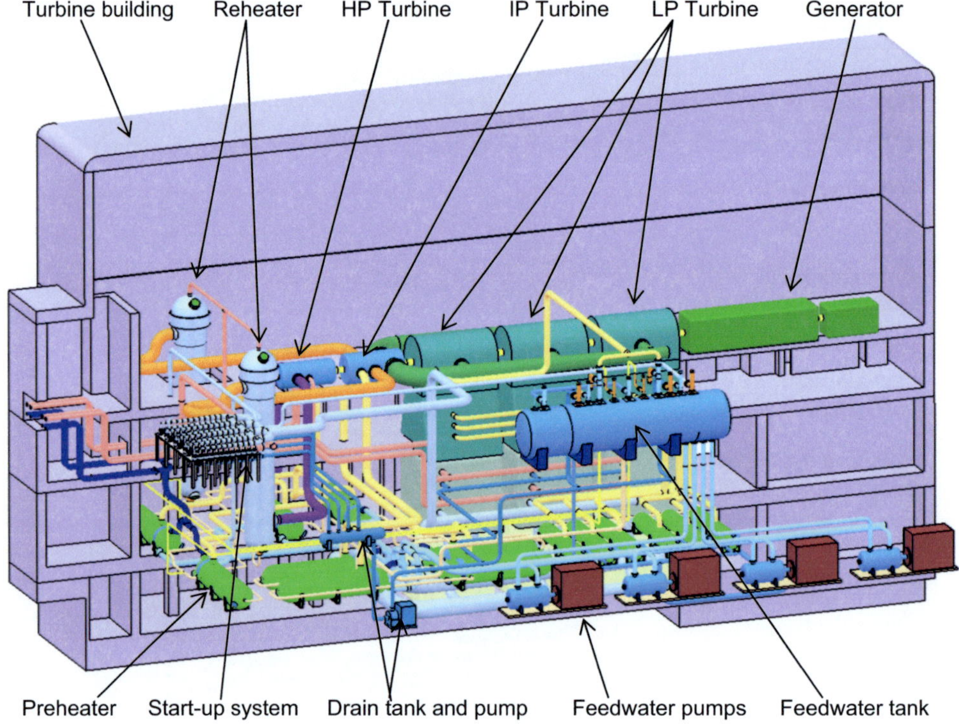

Fig. 5.31: Cut away view of the turbine building with steam cycle components [17]

216

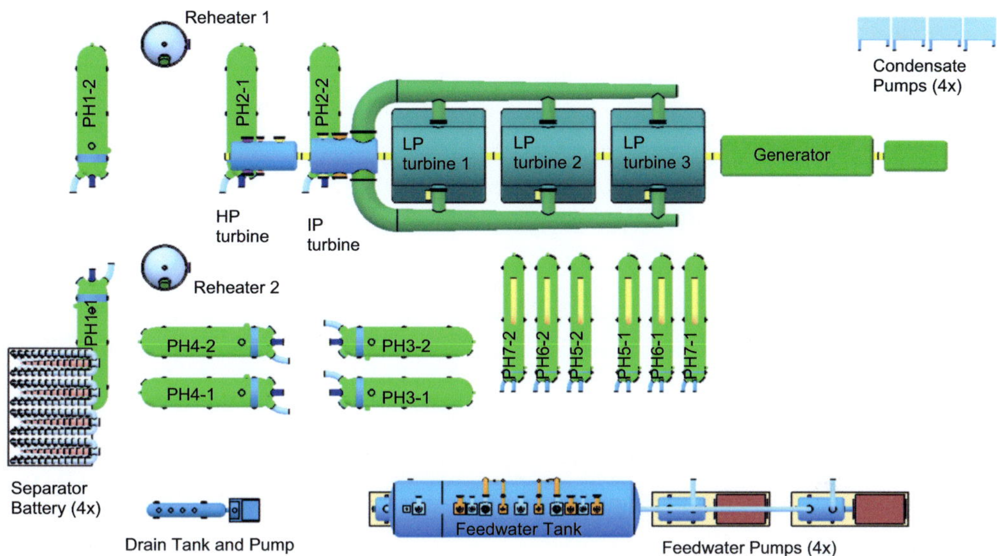

Fig. 5.32: Top view on the arrangement of the main steam cycle components and start-up system [17]

The arrangement of all steam cycle components for the HPLWR is shown in Figure 5.32. The components are placed on different levels of the turbine building and interconnected with pipes. Figure 5.32 shows a top view on the steam cycle components without pipes for better visibility. Because of the different levels of the components, some parts are displayed over each other. The feedwater pumps and the preheaters are installed on the lowest floor, and the turbines, separator batteries and the feedwater tank are on 2 floors above to provide enough pressure head for the drain tank and the feedwater pumps, respectively. A drain pump might be required to pump the liquid from the drain tank to the feedwater tank.

A general plot plan has been proposed in [26] for a seawater site. The plan is based on the general arrangement and size of the buildings of a boiling water reactor considering the size and shape of the reactor and turbine buildings of the HPLWR as described in the previous chapters.

The general characteristics of the plot plan are as follows:

- The reactor building together with the turbine building represents the central complex of the plant.

- Reactor supporting systems building and waste building directly connected to reactor building / turbine building.

- Personnel access to the reactor building is via the controlled access area entrance in the reactor supporting systems building.

- Separate service water intake structures.

- Service water pump buildings and diesel buildings separated by a distance.

The general plot plan is shown in Fig. 5.33. The buildings shown in the plot plan are as follows:

- Reactor building (UJB)

- Reactor containment (UJA)

- Waste building (UKA)

- Reactor supporting systems building (UKB)

- Switchyard (UAA)

- Switchgear building (UBA)

- Offsite system transformer (UBC)

- Auxiliary power transformer (UBE)

- Generator transformer (UBF)

- Emergency diesel generator building (UBP)

- Duct structures (cables)(UBZ)

- Emergency control room building (CB)

- Structure for demineralized water tanks (UGC)

- Vent stack (UKH)

- Turbine building (UMA)

- Duct structure (piping) (UMZ)

- Circulating water intake culvert (UPA)

- Circulating water intake structure (UPC)

- Service water intake structure (UPD)

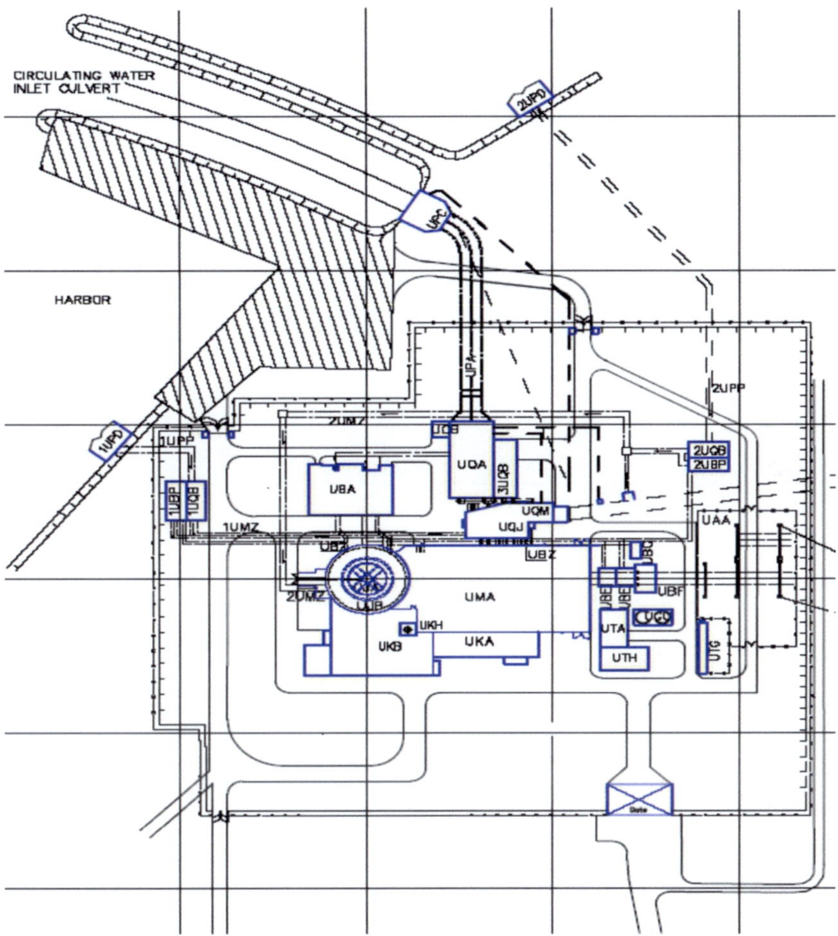

Fig. 5.33: General plot plan of the HPLWR power plant on a seawater site [26]

5.6 Cost estimate and discussion

One of the main objectives of the HPLWR is its economic competiveness thanks to its higher thermal efficiency and its lower plant erection costs compared with conventional PWR or BWR power plants. While the increase of net efficiency has already been shown in Chapter 5.2 by thermodynamic analysis, an estimate of the plant erection costs will need a more detailed discussion as outlined by Starflinger et al. [27]. Let us first look at some qualitative indicators of components costs, as summarized in Tab. 5.4. Two modern nuclear power plants in Germany, the BWR Gundremmingen and the PWR in Neckarwestheim GKN2 [[28], [29]] have been selected here for a coarse comparison with the HPLWR.

		PWR	BWR	HPLWR
Net electric power	MW_e	1400	1344	1000
Steel masses				
- RPV	t	370	785	656
- Closure head	t	116	incl.	122
- Steam generator (SG)	t	490	0	0
- No of SG		4	0	0
- Recirculation Pumps (RP)	t	100	3	0
- No of RP		4	8	0
Total steel mass of the	t	2846	809	778
primary system	t/MW_e	**2.03**	**0.60**	**0.78**
Total volume of the	m^3	65450	22931	9051
containment	m^3/MW_e	**46.75**	**17.06**	**9.05**
Mass of turbine train	t	2860	2860	1430

Table 5.4 Qualitative cost indicators comparing the HPLWR with PWR and BWR plants [27]

The first indicator discussed in [27] is the steel mass on the primary system. For the reference PWR, the main components, reactor pressure vessel (RPV) and its closure head, the four steam generators and the four main circulation pumps have been selected. Summing up the weights, a specific indicator of 2.03 t steel/MW_e has been estimated for these components. For the reference BWR, we get a value of 0.6 t steel/MW_e. There, steam generators are included in the RPV and the internal recirculation pumps are light weighted compared with the PWR. For the HPLWR, we obtain a value of 0.78 t steel/MW_e. The HPLWR has neither a steam generator nor recirculation pumps, but the supercritical pressure requires a larger wall thickness resulting in a total mass of the pressure vessel and closure head of 778 t. Therefore, in this comparison, the BWR has some advantages compared with the HPLWR.

The second cost indicator is the volume of the containment. The volumes were taken from containment drawings, not including all their internal components, and thus provide a maximum value of the total inner volume. The reference PWR has a ratio of 46.75 m^3/MW_e, whereas the reference BWR with 17.06 m^3/MW_e and the HPLWR with 9.05 m^3/MW_e provide smaller values. It must be mentioned here that the comparison of a pressure suppression containment (BWR) with a containment which can be pressurized (PWR) is not really fair, but is shows that a cost reduction had already been included in the development of BWR containments saving concrete. The HPLWR ratio is even smaller than the one of the BWR which shows a significant advantage of this concept. Smaller containment requires less concrete and steel, which will have a positive effect on cost savings.

The third indicator is the ratio of the turbine mass and the electric power. According to Herbell et al. [12], the mass of the HPLWR turbine is about half of the mass of existing referents plants. The resulting cost indicators show also an advantage of the HPLWR turbine (1.43 t/MW_e), which is mainly caused by using a full speed turbine for the HPLWR instead of half speed turbines (being larger) for the reference power plants (PWR: 2.04 t/MW_e; BWR: 2.13 t/MW_e).

This qualitative assessment confirms already that a certain potential can be expected for cost savings of a HPLWR compared with a conventional LWR.

For the estimation of the capital cost, the guidelines of the GIF Economics Working Group [30] recommend to use a top-down model (see Table 5.5) for cost estimation for Generation IV plants, which are in a development state like the HPLWR. In this model, a reference plant has to be defined and the cost differences of the HPLWR to the reference are estimated using available information and data or engineering judgment.

Starflinger et al. [27] compare the plant construction and the electricity generation costs of a HPLWR with a typical LWR. As reference value for the LWR capital costs a value of 3000 $/kWe was taken from actual literature [31]. This value was converted to a reference Euro value of 2200 €/kW_e. According to the Cost Estimating Guidelines for Generation IV Nuclear Energy Systems [30], a specific break down of costs has been executed for both the 1000 MW reference plant and the HPLWR. The cost break-down of the reference plant (see Table 5.5, 3rd column) was taken from a description of the ABWR available on the web [32]. For this analysis it has been assumed that the breakdown of costs can be applied, as a first guess, for the scaled reference plant as well. With the specific costs, the electrical power and the cost breakdown available, the costs within main cost categories (in blue) like "structures and improvements" (account 21), "reactor equipment" (account 22), etc. were calculated.

With the results of the HPLWR available so far, each category has been evaluated regarding possible cost reductions. As displayed in Table 5.5, cost savings can be expected in the order of 20% in all major cost categories. For example, the size reduction of the reactor building and containment reduce the construction costs in account 221 of about 41% (equivalent to about M€ 80). In the entire account 21 "structures and improvements", the construction costs of M€ 430 could be reduced to M€ 334, which equals a saving of 22.4% (M€ 96) compared with the reference value. The cost structure is also visualized in Fig. 5.34.

Adding all direct costs, a saving of M€ 289 (20.4%) could be expected compared with the reference power plant. For the indirect costs, a proportional reduction to the indirect costs (20.4%) was assumed. Finally, the total overnight costs of the HPLWR power plant was calculated to M€ 1,795, 20.4% (M€ 460) less compared to the reference plant which costs M€ 2,255 in this example. The specific plant construction costs amount to 1,795 €/kWe for the HPLWR and 2,255 €/kWe for the reference plant. These data, however, are still having a high uncertainty. Once more information about the design of the HPLWR plant is available, these numbers should be re-evaluated.

Account	Description	Cost breakdown in % of total cost	Costs of 1000 MW reference plant	Fraction HPLWR	HPLWR costs
211	Site preparation	1,5	30.714.286 €	1	30.714.286 €
212	Reactor building	10	204.761.905 €		120.153.933 €
	Containment	5	102.380.952 €	0,431	44.117.787 €
	Structure	5	102.380.952 €	0,743	76.036.145 €
213	Turbine Generator building	2	40.952.381 €	1	40.952.381 €
215	Reactor auxiliary building	2	40.952.381 €	0,8	32.761.905 €
216	Radwaste building	1	20.476.190 €	1	20.476.190 €
218A	Control building	1	20.476.190 €	1	20.476.190 €
218B	Administration building	0,5	10.238.095 €	1	10.238.095 €
218T	Emergency power generation building	2	40.952.381 €	1	40.952.381 €
21xx	Miscellaneous buildings	0,5	10.238.095 €	1	10.238.095 €
21xx	Ultimate heat sink	0,5	10.238.095 €	0,670	6.857.045 €
21	**STRUCTURES AND IMPROVEMENTS**	**21**	**430.000.000 €**		**333.820.501 €**
221	Reactor equipment	10	236.363.636 €		
	reactor vessel	6	141.818.182 €	0,793	112.441.558 €
	control rod drive system	1	23.636.364 €	0,7	16.545.455 €
	reactor internals	3	70.909.091 €	0,565	40.079.051 €
222	Main heat transport system	0,7	16.545.455 €	0	0 €
223	Safety systems	1,7	40.181.818 €	0,7	28.127.273 €
	residual heat removal system				
	safety injection systems				
	containment heat removal				
	combustible gas control				
224	Radioactive Waste processing system	1	23.636.364 €	1	23.636.364 €
225	Fuel handling system	0,8	18.909.091 €	1	18.909.091 €
226	Other reactor plant equipment	2,7	63.818.182 €	1	63.818.182 €
227	Reactor instrumentation and control	4,5	106.363.636 €	1	106.363.636 €
228	Reactor plant miscellaneous	0,6	14.181.818 €	1	14.181.818 €
22	**REACTOR EQUIPMENT**	**22**	**520.000.000 €**		**424.102.428 €**
231	Turbine generator	7,5	143.750.000 €	0,6	86.250.000 €
233	Condensing system	1,2	23.000.000 €	0,7	16.100.000 €
234	Feedwater heating system	1,4	26.833.333 €	1,1	29.516.667 €
235	Other turbine plant equipment	1	19.166.667 €	1	19.166.667 €
236	Instrumentation and control	0,6	11.500.000 €	1	11.500.000 €
237	Turbine plant miscellaneous items	0,3	5.750.000 €	1	5.750.000 €
23	**TURBINE GENERATOR EQUIPMENT**	**12**	**230.000.000 €**		**168.283.333 €**

Table 5.5   Cost distribution of the 1000MW reference plant and the HPLWR [27]

261	Transportation and lift equipment	0,7	9.000.000 €	0,8	7.200.000 €
262	Air, water, plant fuel oil, and steam service systems	1,8	23.142.857 €	0,8	18.514.286 €
263	Communication equipment	0,5	6.428.571 €	1	6.428.571 €
264	Furnishing and fixtures	0,5	6.428.571 €	1	6.428.571 €
26	**MISCELLANEOUS EQUIPMENT**	**3,5**	**45.000.000 €**		**38.571.429 €**
2	**TOTAL DIRECT COST**	**66**	**1.420.000.000 €**		**1.130.077.691 €**
31	Field indirect costs		190.000.000 €		
32	Construction supervision		250.000.000 €		
33	Commissioning and start-up costs				
34	Demonstration test run				
35	Design services offsite				
36	PM/CM services offsite				
37	Design service onsite				
38	PM/CM services onsite				
39	Contingency on support services		125.000.000 €		
	Engineering Home Office		70.000.000 €		
	Owners Costs		200.000.000 €		
30	**CAPITALIZED INDIRECT SERVICES COST**	**34**	**835.000.000 €**	0,7958	**664.517.515 €**
	TOTAL OVERNIGHT COST	**100**	**2.255.000.000 €**		**1.794.595.207 €**
		el Power	1000	el Power	1000
		specific costs	2.255 €	specific cost:	1.795 €

Tab. 5.5, cont.: Cost distribution of the 1000MW reference plant and the HPLWR [27]

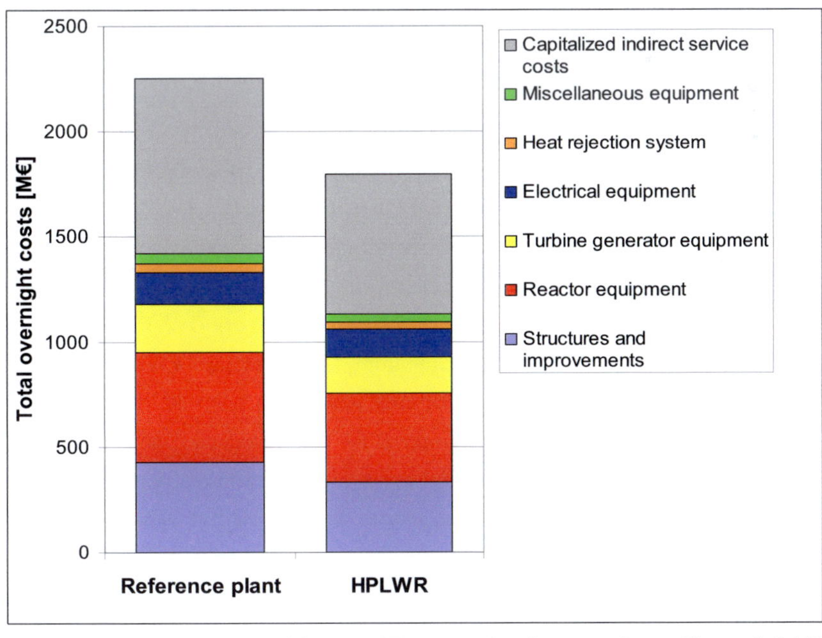

Fig. 5.34: Estimated cost structure of the HPLWR power plant in comparison with a scaled ABWR as reference plant [27]

As shown in Table 5.5, the reduction of the containment size and the higher plant efficiency do not play a dominating role in determining the construction costs. Therefore, it is necessary to continuously assess all accounts, i.e. assess each system and component for possible cost reductions.

It should be mentioned here, that the absolute values in Euro in Table 5.5 are expected to change within time and just provide an adequate, but indicative value for cost saving. Therefore, a parametric study has been carried out to investigate the influence of e.g. construction costs on the electricity generation costs. The following parameters were varied:

- Specific capital costs (10 and 20% less than LWR, and equal to LWR)

- Sensitivity of fuel cycle cost in case of a HPLWR variant with 10 less capital costs than a LWR.

Tables 5.6 and 5.7 show the electricity generation costs at certain stages of operation. Ten and twenty years of operation are located within the depreciation period (25 years), whereas after thirty and forty years of operation, capital costs do not contribute to the electricity generation costs anymore.

Table 5.6 depicts the sensitivity analyses varying the specific plant erection costs within 2200 – 1760 €/kWe. The fuel costs are selected constant to 0.79 €cents/kWh. Within the depreciation period, the capital costs have a certain influence on the costs of electricity (difference optimistic – pessimistic: 0.54 €cent/kWh). Afterwards, the constant fuel price leads to equal electricity generation costs for all different plant erection costs. The electricity generation costs are increasing a little between 30 and 40 years due to the price increase rate assumed (3%).

In Table 5.7, the plant erection costs are held constant to -10% of the reference value and the fuel costs are varied. A high value of 1.04 €cent / kWh and a low value of 0.63 €cent / kWh were selected to investigate raising and falling fuel prices on electricity generation costs. As seen in Table 5.7, the electricity generation costs are lower than the reference case and the spreading for the optimistic and pessimistic fuel costs are 0.38 €cent / kWh after ten years of operation. After twenty years, the spreading is 0.33 €cent / kWh. It should be noted that in this particular case, the HPLWR electricity generation costs with the pessimistic fuel costs (3.57 €cent / kWh) are higher than the costs for the reference case (3.51 €cent / kWh). The reason for this is that the reference fuel price of the reference plant is lower than the pessimistic one of the HPLWR. Five years before the end of the depreciation period, the capital cost still have an influence on the electricity generation costs, but not as decisive as in the beginning. This trend can also be seen after 30 and 40 years of operation in which the electricity generation costs do not change over time and are clearly determined by the fuel costs.

		Reference case	Optimistic	Best estimate	Pessimistic
Plant erection costs	€ / kW$_e$	2200	1760	1980	2200
Fuel costs	€cent / kWh	0.79	0.79	0.79	0.79
Electricity generation costs after	10a	€cent / kWh	4.86	4.32	4.59	4.86
	20a	€cent / kWh	3.51	3.19	3.35	3.51
	30a	€cent / kWh	1.78	1.78	1.78	1.78
	40a	€cent / kWh	1.84	1.84	1.84	1.84

Tab. 5.6: Electricity generation costs after years of operation. Specific fuel costs are constant. Specific capital costs are variable (in blue).

		Reference case	Optimistic	Best estimate	Pessimistic
Plant erection costs	€ / kW$_e$	2200	1980	1980	1980
Fuel costs	€cent / kWh	0.79	0.63	0.79	1.04
Electricity generation costs after	10a	€cent / kWh	4.86	4.48	4.59	4.76
	20a	€cent / kWh	3.51	3.24	3.35	3.57
	30a	€cent / kWh	1.78	1.67	1.78	2.00
	40a	€cent / kWh	1.78	1.67	1.78	2.00

Tab. 5.7: Electricity generation costs after years of operation. Specific capital costs are constant at -10% of reference value. Specific fuel costs are variable (in blue).

As a conclusion, the cost advantage of the HPLWR power plant is plausible, even though the data base is still rather uncertain. Not included in these costs are research and development costs of this novel reactor type, the costs of a prototype and of the first of a kind, which must be paid back by the fleet later on. With this regard, a continous development from existing PWR or BWR in an evolutionary process might be more economic.

5.7   References

[1]   Y. Oka, S. Koshizuka, Y. Ishiwatari, A. Yamaji, Super Light Water Reactors and Super Fast Reactors, Springer, ISBN 978-1-4419-6034-4 (2010)

[2]   M. Pflugbeil, BoA Niederaussem – der 950 MW Braunkohlekraftwerksblock mit optimierter Anlagentechnik, VDI Berichte 1456, P. 133-142, 1999

[3]   R.J. Heitmueller, A. Kather, Thermal and fuel engineering concept of the steam boiler for the BoA unit Niederaussem K, Germany, VGB Kraftwerkstechnik 1999, 79 (5), 75-82

[4]   L. Behnke, S. Himmel. C. Waata, E. Laurien, T. Schulenberg, Prediction of heat transfer for a supercritical water test with a four pin fuel bundle, Proc. ICAPP 06, Paper 6211, Reno, USA, June 4-8, 2006

[5]   L. Behnke, Vorhersage kritischer Siedezustände in Brennelementstrukturen von Leichtwasserreaktoren bei variablem Druck, Dissertation, University of Stuttgart, 2009

[6]   M. Brandauer, M. Schlagenhaufer, T. Schulenberg, Steam cycle optimization for the HPLWR, 4th Int. Symp. on SCWR, Paper 36, Heidelberg, Germany, March 8-11, 2009

[7]   H. Herbell, T. Schulenberg, Reheater design concept for the High Performance Light Water Reactor, Proc. GLOBAL 2009, Paper 9345, Paris, France, Sept. 6-11, 2009

[8]   G.P. Bogoslovskaya, V.M. Abdulkadyrov, Estimation of one-through direct thermodynamic cycle parameters for Supercritical Water Cooled Reactor, Proc. NUTHOS-8, Paper N8P0163, Shanghai, China, Oct. 10-14, 2010

[9]   K. Yamada, S. Sakurai, Y. Asanuma, R. Hamazaki, Y. Ishiwatari, K. Kitoh, Overview of the Japanese SCWR concept developed under the GIF collaboration, 5th Int. Symp. SCWR (ISSCWR-5), Vancouver, Canada, March 13-16, 2011

[10] R.B. Duffey, I. Pioro, X. Zhou, U. Zirn, S. Kuran, H. Khartabil. M. Naidin, Supercritical water-cooled nuclear reactors: current and future concept – steam cycle options, Proc. ICONE 16, Paper 48869, Orlando, USA, May 11-15, 2008

[11] H. Herbell, Thermohydraulische Auslegung des Zwischenüberhitzers eines High Performance Light Water Reactors, Dissertation, Karlsruhe Institute of Technology, 2011

[12] H. Herbell, M. Wechsung, T. Schulenberg, A turbine design concept for the High Performance Light Water Reactor, 4th Int. Symp. on SCWR, Paper 76, Heidelberg, Germany, March 8-11, 2009

[13] H. Herbell, J. Starflinger, T. Schulenberg, Numerical investigation of cooling heat transfer of supercritical water, 5th Int. Symp. SCWR (ISSCWR-5), Vancouver, Canada, March 13-16, 2011

[14] H. Herbell, A. Class, J. Starflinger, T. Schulenberg, Design and stability limits of the HPLWR reheater, Proc. ICAPP 10, Paper 10054, San Diego, USA, June 13-17, 2010

[15] KSB, www.ksb.de/ksb-en

[16] D. Lemasson, Design of a feedwater tank for the High Performance Light Water Reactor, Internship Report, Karlsruhe Institute of Technology, Institute for Nuclear and Energy Technologies, 2009

[17] C. Koehly, J. Starflinger, T. Schulenberg, M. Brandauer, D. Lemasson, R. Velluet, H. Herbell, Draft layout of the HPLWR power plant, 5th Int. Symp. SCWR (ISSCWR-5), Vancouver, Canada, March 13-16, 2011

[18] M. Brandauer, HPLWR Wasser-Dampf-Kreislauf-Optimierung und mechanische Auslegung der Speisewasservorwärmung, Studienarbeit Karlsruhe Institute of Technology, Institute for Nuclear and Energy Technologies, 2009

[19] T. Schlageter, Strukturmechanische Auslegung der Rohrplatte eines Speisewasser-vorwärmers des HPLWR, Diplomarbeit, Karlsruhe Institute of Technology, Institute for Nuclear and Energy Technologies, 2010

[20] M. Schlagenhaufer, J. Starflinger, T. Schulenberg, Steam cycle analyses and control of the HPLWR plant, 4th Int. Symp. on SCWR, Paper 76, Heidelberg, Germany, March 8-11, 2009

[21] APROS Version 5.08, VTT and Fortum, http://apros.vtt.fi

[22] M. Schlagenhaufer, J. Starflinger, T. Schulenberg, Plant control of the High Performance Light Water Reactor, Proc. GLOBAL 2009, Paper 9339, Paris, France, Sept. 6-11, 2009

[23] R. Velluet, Design of the water separator for the combined shut-down/start-up system of the HPLWR, Internship Report, Karlsruhe Institute of Technology, Institute for Nuclear and Energy Technologies, 2009

[24] M. Schlagenhaufer, Simulation des Dampf-Wasserkreislaufs und der Sicherheitssysteme eines High Performance Light Water Reactors, Dissertation, Karlsruhe Institute of Technology, 2010

[25] D. Bittermann, S. Rothschmitt, J. Starflinger, T. Schulenberg, Reactor and turbine building layout of the High Performance Light Water Reactor, Annual Meeting on Nuclear Technology, Paper No. 1216, Berlin, Germany, Mai 4-6, 2010

[26] J. Starflinger, T. Schulenberg, P. Marsault, D. Bittermann, E. Laurien, C. Maraczy, H. Anglart, J.A. Lycklama, M. Andreani, M. Ruzickova, T: Vanttola, A. Kiss, M. Rohde, R. Novotny, High Performance Light Water Reactor Phase 2, Final Public Report, Assessment of the HPLWR Concept, Dec. 2010, www.hplwr.eu

[27] J. Starfinger, T. Schulenberg, D. Bittermann, M. Andreani, A. Toivonen, J.A. Lycklama, Assessment of the High Performance Light Water Reactor, 5th Int. Symp. SCWR (ISSCWR-5), Paper P056, Vancouver, Canada, March 13-16, 2011

[28] Informationen rund um die Kernenergie und die Druckwasserreaktoren des GKN, Gemeinschaftskernkraftwerk Neckarwestheim, 8. Auflage, 2002

[29] Kernkraftwerk Gundremmingen Block B und C, Siemens Energierzeugung, Siemens AG, Erlangen, Bestellnummer A19100-U51-A231-V2, 1995.

[30] Cost estimating guidelines for Generation IV nuclear energy systems, Revision 4.2, September 2007, GIF/EMWG/2007/004

[31] Nucleonics Week, Volume 49/Number 27/July 3, 2008

[32] General description of the ABWR, Chapter 12, "Plant Economics and Project Schedule", Oct. 2010, http://www.ftj.agh.edu.pl/~cetnar/ABWR/Chapter12.pdf

A Annex 1

HPLWR Design Data

Overall Performance

Reactor thermal power (fixed)	2300	MW
Gross Power Output	1046	MW
Net electrical output	1000	MW
Net Plant Efficiency	43.5	%
Plant design life	60	years
Feedwater mass flow rate	1179	kg/s
Feedwater pressure	25	M Pa

Balance of Plant

Steam cycle

Number of feedwater lines at RPV inlet	4	-
Number of steam lines	4	-

Feedwater pumps

Number of feedwater pumps	4 x 33%	-
Nominal mass flow per pump	393	kg/s
Inlet pressure	0.55	M Pa
Outlet pressure	26.7	M Pa
Inlet temperature	155.5	°C
Pump Type	radial	
Pump efficiency	85	%
Pump main dimensions	dia. 2000, L 4000	mm
Mass flow control	speed controlled	

Condensate pumps

Number of condensat pumps	4 x 33%	-
Nominal mass flow per pump	240.3	kg/s
Inlet pressure	4.75	kPa
Outlet pressure	1.35	M Pa
Inlet temperature	31	°C
Pump type	radial	
Pump efficiency	85	%
Pump main dimensions	dia 1000, L 3000	mm
Controlled	head controlled	

Pumps of main heat sink

Number of pumps	3 x 50%	-
Nominal mass flow per pump	14715	kg/s
Inlet pressure	0.1	M Pa
Outlet pressure	0.17	M Pa
Inlet temperature (sea water cooling, cooling tower)	15	°C
Pump type	axial	

Pump efficiency	85	%
Pump main dimensions	dia 2000	mm
Controlled	speed controlled	-
Preheater		
Number of preheater lines	2x50%	
Number of preheaters	7 per line	-
Number of High pressure (HP) preheaters	4	-
Type of HP preheaters	u-tube	-
Material of HP preheaters	ferritic/ martensitic	-
Number of Low pressure (LP) preheaters	3	-
Type of LP preheaters	u-tube	-
Material of LP preheaters	ferritic/ martensitic	-
Pressure drop (only friction in tube-side (P91 as material))		
PH 1	0.03	M Pa
PH 2	0.02	M Pa
PH 3	0.04	M Pa
PH 4	0.04	M Pa
PH 5	0.05	M Pa
PH 6	0.02	M Pa
PH 7	0.03	M Pa
Feedwater tanks		
Number of feedwater tanks	1	-
Nominal pressure	0.55	M Pa
Nominal temperature	155.5	°C
Total tank volume	375	m³
Therefrom water volume	187.5	m³
Material	ferritic/ martensitic	-
Turbines		
Mechanical efficiency	0.994	-
HP turbine		
Number of HP turbines	1 x 2 flooded	-
Isentropic efficiency	0.88	-
Number of inlet pipes	4	-
Inlet pressure	22.6	M Pa
Inlet temperature	494	°C
Number of extractions	1	-
IP turbine		
Number of IP turbines	1 x 2 flooded	-
Isentropic efficiency	0.94	

Number of inlet pipes	2	-
Inlet pressure	4.04	M Pa
Inlet temperature	441	°C
Number of extractions	2	-
LP turbine		
Number of LP turbines	3 x 2 flooded	-
Isentropic efficiency	0.84	
Number of inlet pipes	3 x 2 pipes	-
Inlet pressure	0.78	M Pa
Inlet temperature	233	°C
Outlet pressure	5	kPa
Steam quality (at Outlet)	0.87	-
Reheater		
Number of reheaters	2 x 50%	-
Inlet pressure (shell side)	4.25	M Pa
Inlet temperature (shell side); steam quality>1	260.2	°C
Inlet pressure (pipe side)	22.6	M Pa
Inlet temperature (pipe side)	494	°C
Pressure drop (shell side)	0.2	M Pa
Reheat temperature	435	°C
Condenser		
Number of condensers	3	-
Condenser pressure	5	kPa
Pressure drop	0.25	kPa
Main heat sink		
Inlet temperature	15	°C
Inlet pressure	0.1	M Pa
Maximum outlet temperature	25	°C
Generator		
Generator Type	THDF - SIEMENS	
Electrical efficiency	0.99	-
Mechanical efficiency	0.99	-

Nuclear island (inside containment)

Reactor Coolant System (RCS)		
RCS operating pressure (reactor inlet)	25	M Pa
RCS design pressure (115%)	28.75	M Pa
Vessel inlet temperature	280	°C
Vessel outlet temperature	500	°C
Vessel pressure drop		M Pa
Total flow rate	1179	kg/s
Flow split in RPV:		
-Downcomer	50%	
-Uppler Plenum	50%	
-Gap water	50%	
-Moderator water	50%	
Containment		
Maximum design pressure	0.5	M Pa
Operating pressure	0.1	M Pa
Drywell atmosphere	Nitrogen	
Dry well gas volume	2131	m³
Total water volume	2021	m³
Containment inner diameter	20	m
Containment inner height	23.7	m
Valves		
No of Main Steam Line Isolation Valves (MSIV)	4	-
Types	e.g. Sempell type 614-324 and EBS 32	
Stroke time	3	s
Actuation	passive and active	
No of Main Feedwater Checkvalves	4	-
Type	e.g. Sempell KR-400	
No. of Safety relief valves	8	-
Cross Section Area	110	cm2
Type	e.g. Sempell VSH	
Actuation	passive and active	
Actuation pressure (110% of operation pressure)	27.5	M Pa
Overpressure Protection		
Automatic depressurization system (ADS)	4 discharge trains; each train provided with two safety/relief valves	
	Discharge flow to core flooding pool via 8 spargers	
Core flooding pool		
Arrangement	inside containment	-
Number of pools	4	-

Total water volume	1121	m³
Initial pool temperature	40	°C

Pressure suppression pool (PSP)

Arrangement	inside containment	-
Number of pools	1	-
Total water volume	900	m³
Gas volume (nitrogen)	500	m³
Max. pool temperature	80	°C
Initial pool temperature	40	°C

Active Residual Heat Removal System (RHRS) (LPCI)

Arrangement	inside containment	-
Number of pumps	4 x 100%	-
Actuation pressure	6	M Pa
Inlet pressure	0.1	M Pa
Outlet pressure	6	M Pa
Mass Flow	180	kg/s
Type	radial	
Material of pump	SS316 L (N)	-
Pump efficiency	85	%
Pump main dimensions	dia.800, length 2000	mmm
Controlled	constant speed	-
Number of heat exchangers (HX)	4	-
Tube side inlet temperature for HX	87	°C
Tube side outlet temperature from HX	57	°C
Shell side inlet temperature	37	°C
Shell side outlet temperture	63	°C
Mass flow tube side	140	kg/s
Mass flow shell side	162	kg/s
Pressure drop tube and shell side	<0.1	M Pa
Material heat exchanger	SS316 L (N)	-
Connecting piping: inner diameter	200	mm
Connecting piping: outer diameter	210	mm
Tube bundle length	ca. 3000	mm
Overall length	ca. 5000	mm
Diameter	ca. 1500	mm
Piping material	SS316 L (N)	-
Total number of check valves	4	-

Passive RHRS (HPCI)

to be designed	

Containment Heat Removal System

Long term containment pressure control	4 containment cooling condensers

Poisoning System		
Boron concentration	20 to 25 % B10	%
Number of trains and pumps	2	
Safety functions	Secondary shut-down system	
Volume of borated water	ca. 10	m³
H2 Overflow pipe		
Number of overflow pipes	2	-
Diameter of overflow pipes	600	mm
Depth in PSP	to be determined	mm
Elevation	to be determined	mm
Core flooding pool overflow pipe to PSP		
Number of overflow pipes	4	-
Diameter of overflow pipes	400	mm
Vent pipes (between Core flooding pool and PSP)		
Number of pipes	16	-
Inner diameter of pipes	600	mm
Emergency power supply	Emergency diesels and SBO diesels	

Reactor Pressure Vessel and Internals

Reactor Pressure Vessel (RPV) and Closure Head

Design temperature	350	°C
Height (incl. closure head)	14.29	m
Inner diameter of RPV	4464.7	mm
Wall thickness (cyl. shell)	446	mm
Wall thickness (bottom head)	300	mm
Wall thickness (upper flange)	558	mm
Wall thickness (closure head)	400	mm
Number of nuts and bolts	40 x M210x8	-
Material vessel and closure head	20 MnMoNi 5 5	-
Material nuts and bolts	Property class 12.9	-
Weight of RPV	656124	kg
Weight of closure head, nuts, bolts and sealing	121755	kg
Reactor gas volume (empty, without internals)	199.6	m³
Reactor gas volume (assembled RPV with internals)	150.2	m³

Upper Mixing Chamber/ Steam plenum

Assembly of base and separation plates, jacket, stiffening tubes and mixing walls		
Design temperature	500	°C
Outer diameter of steam plenum (without support brackets)	3970.7	mm
Outer diameter of steam plenum support brackets	4150.7	mm
Max. inner diameter of steam plenum	3910.7	mm
Height of upper mixing plenum (inside)	480	mm
Wall thickness (horizontal plate)	60	mm
Wall thickness (peripheral shell)	30	mm
Diameter of upper head piece opening	214,5	mm
Diameter of lower head piece opening	222.7	mm
Material of steam plenum	1.4970	-
Weight of steam plenum	8918	kg

Lower Mixing Chamber/ Lower Mixing Plenum

Assembly of core base plate, mixing chamber and half shell mixing wall		
Design temperature	500	°C
Max. outer diameter of lower mixing chamber (flange)	3970.7	mm
Min. outer diameter of lower mixing chamber	3850.7	mm
Max. inner diameter of mixing chamber zone 2	3608.6	mm
Height of lower mixing plenum (inside)	550	mm
Wall thickness (core base plate)	300	mm
Wall thickness (mixing chamber base plate)	25	mm
Wall thickness (mixing chamber side walls)	20	mm
Diameter (foot piece opening / core base plate)	150	mm
Number of foot piece openings (core base plate)	156	-

Diameter of steel reflector water outlets in core base plate	70	mm
Number of steel reflector water outlets in core base plate	88	-
Material of lower mixing chamber	1.4970	-
Weight of lower mixing chamber (with orifices and swirlers)	27493	kg
Spring element		
between core barrel and control rod guiding tubes		
Design temperature	350	°C
Inner diameter of spring element	4454.7	mm
Outer diameter of spring element	4584.7	mm
Material of spring element	Inconel 718	-
Weight of spring element	368	kg
Control rod (CR) guiding tubes		
Assembly of 156 guiding tubes connected in 2 base plates		
Design temperature	350	°C
Outer diameter of CR guiding tubes (lower plate)	3970.7	mm
Outer diameter of CR guiding tubes (upper flange plate)	4664.7	mm
Height of CR guiding tubes	4335.8	mm
Inner diameter of CR guiding tubes	192	mm
Number of CR guiding tubes	156	-
Inner diameter of holes for moderator water	100	mm
Number of coolant openings for gap and moderator water	129	-
Material of CR guiding tubes	1.4970	-
Weight of CR guiding tubes	53120	kg
Core barrel		
Design temperature	500	°C
Inner diameter of core barrel	3990.7	mm
Thickness of core barrel	60	mm
Outer diameter of core barrel	4110.7	mm
Outer diameter of core barrel (flange)	4664.7	mm
Material of core barrel	1.4970	-
Weight of core barrel	61184	kg
Steel reflector with bellow		
Design temperature	500	°C
Mass flow rate for reflector cooling	589.5	kg/s
Height of steel reflector (without bellow)	5200	mm
Water layer thickness	~100	mm
Thickness of steel reflector plates (innner and outer wall)	50	mm
Thickness of filling plates	30	mm
Outer diameter of steel reflector	3990.7	mm
Inner diameter of bellow	3760	mm
Outer diameter of bellow	3950	mm

Material of steel reflector	SS 316 L (N)	-
Material of bellow	1.4541	-
Weight of steel reflector	68846	kg
Weight of bellow	87	kg
Inlet pipe assembly with Backflow Limiter		
Design temperature	350	°C
Number of inlet pipes and backflow limiter	4	-
Inner diameter of inlet pipes (cold legs)	200	mm
Outside Outer diameter of inlet pipes (cold legs)	312	mm
Material of inlet pipes (cold legs)	1.4970	-
Inner diameter of backflow limiter	200	mm
Outside Outer diameter of backflow limiter	230	mm
Material of backflow limiter	1.4970	-
Outlet pipe assembly		
Assembly of hot steam and outlet pipe, hydraulic positioning unit, nuts and bolts		
Design temperature	500	°C
Number of outlet pipes assemblies	4	-
Inner diameter of hot legs	350	mm
Outer diameter of hot legs	412	mm
Inner diameter of outlet pipes	800	mm
Outer diameter of outlet pipes	951	mm
Material of outlet pipes	P91	
Outer diameter of hot steam pipe (circular plates)	795	mm
Outer diameter of flow plates (hot steam pipe)	785	mm
Inner diameter of hot steam pipe	390	mm
Material of hot steam pipes, hydr. positioning unit	1.4970	-
Number of nuts and bolts for outlet flange	20 x M72x6	-
Material of nuts and bolts for outlet flange	Property class 12.9	-
Total weight of 4 outlet pipes assemblies	40650	kg
Length to MSIV	7.4	m
Length to ADS valves	11	m
Length to Turbine bypass valves	68	m

Core and Fuel Assemblies

Core Design

Number of fuel assemblies (FA)	1404	-
Type of FA	Cluster bundles of 9 FA with 40 fuel pins each	
Number of clusters in evaporator, superheater 1, superheater 2	52 each; 156 in total	-
Total core height (plenum to plenum)	5331	mm
Active length	4200	mm
Total FA length (without cluster spring)	6175	mm
Total FA length (incl. cluster spring)	6280	mm
Linear heat rate, nominal	< 390	W/cm
Enrichment	up to 9%	% U 235
Max. discharge burnup (target value)	60	MWd/kg

Cluster

Number of assemblies per cluster	9	-
Gap between assemblies	9	mm
Cluster size incl. gap (maximum)	247.6	mm

Assembly boxes

Assembly box inner side length	67.52	mm
Assembly box wall thickness	3	mm
Assembly box outer size	73.52	mm
Assembly box inner corner radius	2.5	mm
Assembly box outer corner radius	5.5	mm
Assembly box axial length (total height)	4866	mm
Assembly box axial length (only straight part)	4802	mm
Assembly box upper inner diameter	59	mm
Moderator box outer side length	26.88	mm
Moderator box wall thickness	2	mm
Moderator box inner side length	22.88	mm
Moderator box outer corner radius	3.5	mm
Moderator box inner corner radius	1.5	mm
Moderator box outlet nozzle outer diameter	24	mm
Moderator box outlet nozzle inner diameter	20	mm
Material assembly and moderator box	SS 347 honeycomb filled with ZrO_2	
Thickness inner sheet of assembly box	0.6	mm
Thickness all other sheets	0.4	mm
Diameter of venting holes in colder sheets	0.5	mm
Honeycomb structure wall thickness	0.2	mm

Head/ Foot pieces

Inner diameter of head piece (minimum)	199.7	mm

Total height of head piece (incl. Window and bush)	963.44	mm
Inner diameter of foot piece (minimum)	150	mm
Total height of foot piece	390	mm
Material of FA head and foot piece	1.4970	-
Control rods		
Type of CR	square rods, two layers with absorber material in between	
Number	5 CRs joined on 1 con-rod per FA cluster; 156 x 5 in total	
Outer diameter of CR (maximum)	195.1	mm
Diameter of con-rod drive	50	mm
CR box inner side length	13.88	mm
CR box inner and outer wall thickness	1	mm
CR box absorber wall thickness	01. Mai	mm
CR box outer side length	20.88	mm
CR box outer corner radius	1	mm
Material CR boxes	1.4970	-
Material CR absorber	B4C	-
Insertion Time	3.5	s
Concentration of Boron 10	natural	
Fuel pins		
Cladding inner diameter	7	mm
Cladding thickness	0.5	mm
Cladding outer diameter	8	mm
Cladding alloy	PM2000	-
Ratio Pitch / Cladding outer diameter	1.18	-
Pitch	9.44	mm
Diameter wires	1,34	mm
Axial pitch of wire wraps	200	mm
Fuel	UO_2	-
Fuel pellet diameter	6.7	mm
Active height	4200	mm
Upper fission gas plenum	255	mm
Lower fission gas plenum	255	mm
Total height of fuel pins	4760	mm
Wire material	PM2000	
Wire direction	anticlockwise in flow direction of EVA	
Orifices		
Orifices for mass flow distribution		
Lower mixing plenum (inlets in half shell of lower mixing chamber, from downcomer)		
- Number of inlets	13	-
- Diameter of inlets	100	mm

Upper openings in steel reflector (inlet of reflector water)		
- Number of inlets	72	-
- Diameter of inlets	80	mm
FA foot piece openings (outlet of moderator water)		
- Number of outlets	16	-
- Diameter of outlets	16	mm
Evaporator orifices in core base plate of lower mixing chamber (inlet in each evaporator cluster)		
- Number of inlets	52	-
- Inner diameter of inlets	120	mm
Moderator inlets in each FA head piece		
- Number of circular inlet orifices for FA in edges	4	-
- Inner diameter of circular inlet orifice for FA in edges	11	mm
- Number of inlet gaps (between moderator box and CR)	5	-
- Flow cross section of one inlet gap	86.45	mm²
- Number of circular inlet gaps in CR for outer FAs	4	-
- Inner diameter of circular inlet gaps in CR for outer FAs	3	mm
- Number of inlet gaps in CR for central FA	4	-
- Inner diameter of inlet gaps in CR for central FA	2	mm
Orifices for flow stability		
Evaporator orifices in FA foot piece (inlet in each evaporator assembly box)		
- Number of inlets	9	-
- Inner diameter of inlets	38.45	mm
- Outer diameter of moderator box outlet	24	mm